Micro-Resonators

Micro-Resonators

The Quest for Superior Performance

Special Issue Editor

Reza Abdolvand

MDPI • Basel • Beijing • Wuhan • Barcelona • Belgrade

MDPI

Special Issue Editor
Reza Abdolvand
University of Central Florida
USA

Editorial Office
MDPI
St. Alban-Anlage 66
4052 Basel, Switzerland

This is a reprint of articles from the Special Issue published online in the open access journal *Micromachines* (ISSN 2072-666X) from 2017 to 2018 (available at: https://www.mdpi.com/journal/micromachines/special_issues/micro_resonators)

For citation purposes, cite each article independently as indicated on the article page online and as indicated below:

LastName, A.A.; LastName, B.B.; LastName, C.C. Article Title. *Journal Name* **Year**, *Article Number*, Page Range.

ISBN 978-3-03897-626-4 (Pbk)
ISBN 978-3-03897-627-1 (PDF)

Cover image courtesy of Muhammad Wajih Ullah Siddiqi and Joshua E.-Y. Lee.

Contents

About the Special Issue Editor

Reza Abdolvand, is an associate professor and the director of the Dynamic Microsystems Lab in the Department of Electrical Engineering and Computer Science at the University of Central Florida, where he joined in January of 2014. Prior to that he was an assistant professor in the School of Electrical and Computer Engineering at Oklahoma State University after receiving his Ph.D. degree from the School of Electrical and Computer Engineering, Georgia Institute of Technology, Atlanta, GA in 2008. His research interests are in the general area of micro/nanoelectromechanical systems with more than 15 years of experience in design, fabrication, and characterization of micro-resonators. Dr. Abdolvand has authored and co-authored two book chapters and more than 70 peer-reviewed journal and conference articles in his field of expertise. He has been awarded 12 US patents and was a recipient of the National Aeronautics and Space Administration (NASA) Patent Application Award in 2009.

micromachines

MDPI

Editorial

Editorial for the Special Issue on Micro-Resonators: The Quest for Superior Performance

Reza Abdolvand

Dynamic Microsystems Lab, Department of Electrical and Computer Engineering, University of Central Florida, Orlando, FL 32816, USA; reza@ece.ucf.edu

Received: 6 November 2018; Accepted: 14 November 2018; Published: 27 November 2018

Micro-resonators have reached a distinctive level of maturity due to the accumulated wealth of knowledge on their design, modeling, and manufacturing during the past few decades [1]. Alongside this tremendous scientific progress, micro-resonators are now commonly found in most electronic systems. In this Special Issue, our attempt was to look deeper into less-common topics in this field, such as the nonlinear operation of micro-resonators that are envisioned to play a more important role with the evolution of this technological area.

As the energy density in a resonant device increases, the nonlinear effects could no longer be avoided or ignored. Therefore, it is critical to identify and carefully study the system parameters that impact nonlinearity in micro-resonators and to investigate the effect of nonlinearity in the performance. In Reference [2], the authors study how non-idealities, such as etch profile, in the fabrication of capacitive micro-resonators could affect nonlinear behavior of the device, and in Reference [3], a more accurate one-degree-of-freedom model is developed for the prediction of nonlinear behavior in capacitive beam resonators. Furthermore, in Reference [4], a novel resonator design is proposed to excite a 2:1 internal resonance through nonlinear coupling and to study the effect of air-damping loss on the operation of such devices.

Furthermore, the tuning range, frequency stability, and quality factor (Q) of micro-resonators are the focus in References [5–7] correspondingly, all of which are of significant practical importance, specifically in oscillator applications. The authors of [5] propose two methods for extending tuning range through stiffness alteration that could be effectively implemented in torsional resonators. In Reference [6], frequency stability in response to applied acceleration is investigated in bulk-extensional single crystalline silicon resonators and the dependency of acceleration-sensitivity on the resonator orientation with respect to the silicon crystalline planes are studied through finite element modeling and demonstrated through measurement. In Reference [7], the authors present the effectiveness of phononic crystal band-gap structures in improving the Q in bulk-extensional micro-resonators by reflecting acoustic energy back to the acoustic cavity, as they are strategically placed outside the anchors.

Finally, three unconventional micro-resonator structures are explored in References [8–10]. In Reference [8], the authors introduce a technique called chemical foaming to form glass bubbles that could be utilized for the implementation of hemispherical resonators. In Reference [9], an LC tank is presented with a significant size/performance enhancement achieved through the insertion of a coupling capacitance at the center of an air-bridged circular spiral inductor. Lastly, in Reference [10], the authors propose a unique approach to the realization of electromagnetically induced transparency (EIT) through cascaded multi-mode optical micro-ring resonators.

At the end of this brief introduction to the Special Issue, we would like to thank the authors who entrusted us with the publication of their scientific contributions and acknowledge the many expert reviewers whose technical insight has been instrumental in the timely evaluation of the submitted papers.

Conflicts of Interest: The author declares no conflicts of interest.

References

1. Abdolvand, R.; Bahreyni, B.; Lee, J.E.-Y.; Nabki, F. Micromachined Resonators: A Review. *Micromachines* **2016**, *7*, 160. [CrossRef] [PubMed]
2. Feng, J.; Liu, C.; Zhang, W.; Hao, S. Static and Dynamic Mechanical Behaviors of Electrostatic MEMS Resonator with Surface Processing Error. *Micromachines* **2018**, *9*, 34. [CrossRef] [PubMed]
3. Li, L.; Zhang, Q.; Wang, W.; Han, J. Monostable Dynamic Analysis of Microbeam-Based Resonators via an Improved One Degree of Freedom Model. *Micromachines* **2018**, *9*, 89. [CrossRef] [PubMed]
4. Noori, N.; Sarrafan, A.; Golnaraghi, F.; Bahreyni, B. Utilization of 2:1 Internal Resonance in Microsystems. *Micromachines* **2018**, *9*, 448. [CrossRef] [PubMed]
5. Lee, J.-I.; Jeong, B.; Park, S.; Eun, Y.; Kim, J. Micromachined Resonant Frequency Tuning Unit for Torsional Resonator. *Micromachines* **2017**, *8*, 342. [CrossRef] [PubMed]
6. Khazaeili, B.; Gonzales, J.; Abdolvand, R. Acceleration Sensitivity in Bulk-Extensional Mode, Silicon-Based MEMS Oscillators. *Micromachines* **2018**, *9*, 233. [CrossRef] [PubMed]
7. Siddiqi, M.W.U.; Lee, J.E.-Y. Wide Acoustic Bandgap Solid Disk-Shaped Phononic Crystal Anchoring Boundaries for Enhancing Quality Factor in AlN-on-Si MEMS Resonators. *Micromachines* **2018**, *9*, 413. [CrossRef] [PubMed]
8. Xie, J.; Chen, L.; Xie, H.; Zhou, J.; Liu, G. The Application of Chemical Foaming Method in the Fabrication of Micro Glass Hemisphere Resonator. *Micromachines* **2018**, *9*, 42.
9. Kim, E.S.; Kim, N.Y. Micro-Fabricated Resonator Based on Inscribing a Meandered-Line Coupling Capacitor in an Air-Bridged Circular Spiral Inductor. *Micromachines* **2018**, *9*, 294. [CrossRef] [PubMed]
10. Le, T.-T. Electromagnetically Induced Transparency (EIT) Like Transmission Based on 3 × 3 Cascaded Multimode Interference Resonators. *Micromachines* **2018**, *9*, 417. [CrossRef] [PubMed]

micromachines

MDPI

Article

Micromachined Resonant Frequency Tuning Unit for Torsional Resonator

Jae-Ik Lee [†], Bongwon Jeong, Sunwoo Park, Youngkee Eun [‡] and Jongbaeg Kim *

School of Mechanical Engineering, Yonsei University, 50 Yonsei-ro, Seoul 03722, Korea;
sbrm@yonsei.ac.kr (J.-I.L.); jeongbw@gmail.com (B.J.); acidor@gmail.com (S.P.); yeun@yonsei.ac.kr (Y.E.)
* Correspondence: kimjb@yonsei.ac.kr; Tel.: +82-2-2123-2812
† Present address: Department of Ophthalmology, Henry Ford Health System, Detroit, MI 48202, USA.
‡ Present address: Korea Institute of Industrial Technology (KITECH), 143 Hanggaul-ro, Ansan 15588, Korea.

Received: 22 October 2017; Accepted: 22 November 2017; Published: 25 November 2017

Abstract: Achieving the desired resonant frequency of resonators has been an important issue, since it determines their performance. This paper presents the design and analysis of two concepts for the resonant frequency tuning of resonators. The proposed methods are based on the stiffness alteration of the springs by geometrical modification (shaft-widening) or by mechanical restriction (shaft-holding) using micromachined frequency tuning units. Our designs have advantages in (1) reversible and repetitive tuning; (2) decoupled control over the amplitude of the resonator and the tuning ratio; and (3) a wide range of applications including torsional resonators. The ability to tune the frequency by both methods is predicted by finite element analysis (FEA) and experimentally verified on a torsional resonator driven by an electrostatic actuator. The tuning units and resonators are fabricated on a double silicon-on-insulator (DSOI) wafer to electrically insulate the resonator from the tuning units. The shaft-widening type and shaft-holding type exhibit a maximum tuning ratio of 5.29% and 10.7%, respectively.

Keywords: resonant frequency tuning; shaft-widening; shaft-holding; torsional resonator

---

1. Introduction

Micromachined resonators have a broad range of applications owing to their various advantages such as fast response, high sensitivity, small size, low power consumption, and low fabrication cost [1–3]. To ensure high and uniform performance, the micromachined resonators are supposed to have desired resonant frequencies. However, the resonant frequencies of the micromachined resonators often deviate from the intended value. One major source of this deviation is a dimensional error inevitably existing in fabrication process. It has been reported that relative tolerance in microfabrication could be as high as ±20% of the minimum feature size [4]. In addition, the cross section of fabricated microstructures might be trapezoidal instead of intended square shape due to an imperfect ion-reactive etching process [5]. The Microfabrication Laboratory at UC-Berkeley statically analyzed the resonant frequencies of 31 micromachined resonators, and reported a maximum discrepancy up to 4.9% [6]. It is also reported that the micromachined resonators generally have a deviation in resonant frequency between ±1%~±5% [7]. Operation environments such as pressure and temperature can also shift the resonant frequency of the resonators [8,9]. While the fabrication errors are constant after manufacture, environmental factors vary over time, which may cause frequent or continuous changes in resonant frequency.

Previous works for resonant frequency tuning can be categorized into two groups. The first approach is based on the permanent structural modification of the resonators, which is usually referred to as passive frequency tuning [6,10–12]. Passive frequency tuning can be accomplished by increasing the vibrating mass by the post-fabrication process such as pulsed laser deposition [6] and platinum deposition using a focused ion beam (FIB) [10]. Stiffness adjustment has also been presented by using polysilicon deposition [11] and FIB-machining [12]. The main advantage of the passive frequency methods is that they do not consume power to maintain the tuned state. On the other hand, the second approach does not rely on permanent structural modification, and therefore offers reversible and active tuning capabilities [13–23]. Previous works for active frequency tuning utilized the electrostatic spring effect [14,15], the geometry of the capacitors [16–21], the thermal stressing effect [22], and thermal expansion [23]. Only a few works have been reported regarding the frequency tuning of torsional resonators [14,24,25], such as designing angle limiters near the torsional spring [24] or inducing stress on bending flexures [25].

In this paper, we present two concepts for the resonant frequency tuning of torsionally-driven resonators [26]. The proposed methods are based on the micromachined frequency tuning unit, which is integrated together with electrostatic torsional resonator on the same chip. The separation of actuators for the tuning unit and resonator enables independent control over the amplitude and resonant frequency of the resonator, allowing continuous and repetitive tuning.

2. Design and Principle

Our concept for resonant frequency tuning is based on the stiffness alteration of the torsional shaft either by widening the angle of tilted shafts, or by constraining the torsional motion of the shaft. Hereafter, the former and the latter are referred to as shaft-widening and shaft-holding, respectively. The schematics of the frequency tuning units integrated with torsional resonators are shown in Figure 1: (a) shaft-widening type, and (b) shaft-holding type. A staggered vertical comb (SVC) is utilized to actuate the torsional resonator, and a micromirror is designed to optically read the torsional angle of the resonator (Figure 1c). The resonator is torsionally-driven with respect to the torsional shafts by electrostatic force between two sets of fixed and movable comb fingers at different vertical positions, as shown in the inset of Figure 1c. The torsional resonator actuated by SVC sets is fabricated on two different levels of single crystalline silicon layers of double silicon-on-insulator (DSOI) wafer with fixed combs on upper layers (colored in white) and moving combs and mirrors on the lower layer (colored in green). The usage of silicon double layers allows self-alignment between the fixed and moving combs with simple fabrication [27].

The frequency tuning units for shaft-widening are composed of a chevron thermal actuator [28] and a scissor mechanism part. The right ends of the scissor mechanism part (scissor tips) are connected to the tilted shafts of the torsional resonator (Figure 2a). The force from the chevron thermal actuator is used to drive the scissor mechanism. The scissor mechanism with three pairs of compliant segments is designed to amplify the stroke of the thermal actuator and to convert a linear displacement to a scissor blade-like symmetrical angular motion. As depicted in Figure 2a, when the shuttle is pulled leftward by the chevron thermal actuator, the scissor mechanism part is opened at the right end, widening the angle between the tilted shafts. This structural change increases the effective stiffness of the tilted shafts, resulting in an increase of resonant frequency.

The shaft-holding type also employed a chevron thermal actuator and the scissor mechanism, but the shaft-holders are attached to the scissor tips (Figure 2b). The chevron thermal actuator of the shaft-holding type is designed to push the shuttle rightward, as depicted in Figure 2b. Then, the scissor tips are closed at the right end, moving the shaft holders towards the torsional shaft of the resonator. The shaft holders are structured as a bow-like shape, by which the level of restriction upon the torsional spring would be continuously augmented as the shaft-holding flexure exhibits elastic deformation. When the mechanical restriction is initiated, the contact occurs at the middle of the torsional shaft resulting in a relatively lower tuning ratio (Figure 3a). As the shaft-holder moves further, full contact can be formed between the torsional spring and the holding flexure, extending the contact area to the segment close to the mirror (Figure 3b). In this way, both a high tuning ratio and a wide tuning range would be possible.

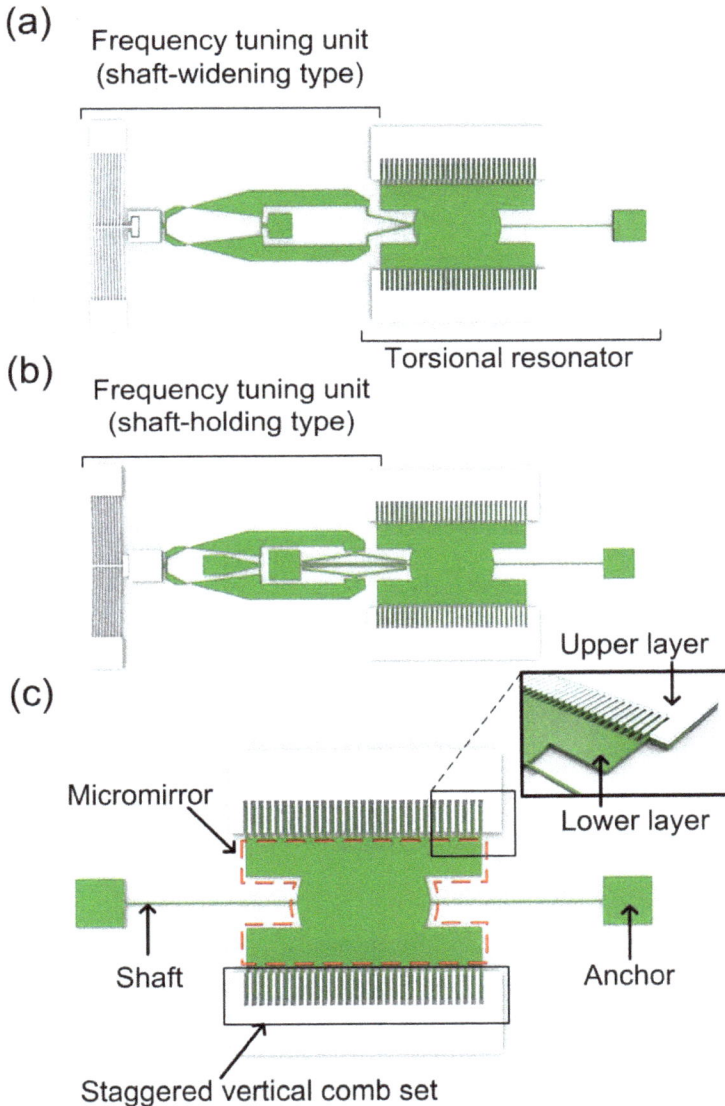

Figure 1. Schematics of frequency tuning units integrated with torsional resonators: (**a**) Shaft-widening type and (**b**) shaft-holding type. (**c**) Schematic layout of staggered vertical comb (SVC)-based torsional resonator. The micromirror is designed to optically read the torsional angle of the resonator. (Inset: close-up view of the SVC set defined on different silicon layers. Upper and lower layers are colored in white and green, respectively). Reproduced with permission from [26].

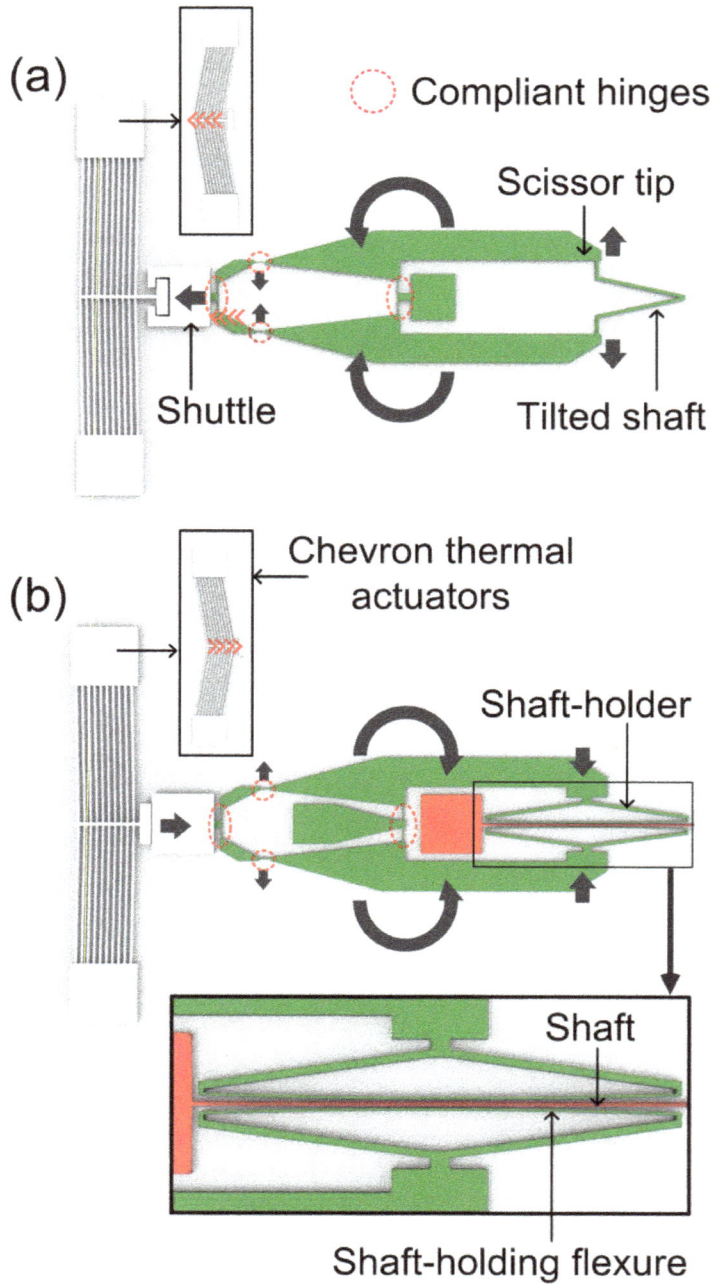

Figure 2. Design of (**a**) shaft-widening and (**b**) shaft-holding type frequency tuning units. Linear motion from the chevron thermal actuator (inset figure) is transformed and amplified to widen the gap between the tilted shafts (shaft-widening type), or to mechanically restrict the rotational motion of the shaft (shaft-holding type).

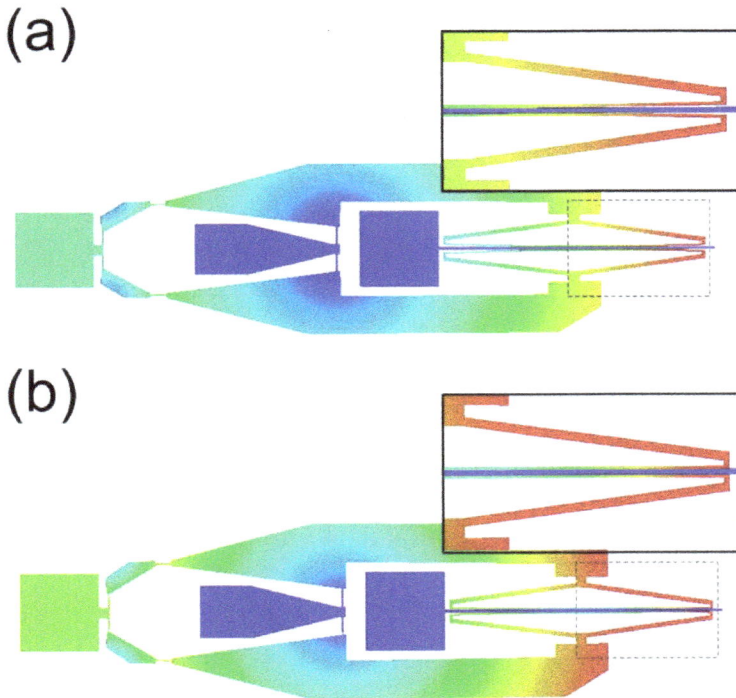

Figure 3. Schematic view of shaft-holding mechanism. (**a**) By the driving force from the chevron thermal actuator, the contact between the shaft-holding flexures and the torsional shaft starts to be formed from the middle of the torsional shaft; (**b**) As the shaft-holder moves further, the shaft-holding flexure is elastically deformed, resulting in a gradual increase in contact area.

One of critical aspects that we considered for the design of the frequency tuning unit is the insulation of the electrical current, which may flow from the chevron thermal actuator to the electrostatically driven torsional resonator, resulting in a malfunction of the resonator. Preventing this charge leakage is another reason for employing the DSOI wafer with two isolated structural silicon layers, as shown in Figure 4. The thermal actuator is formed on the upper layer of the DSOI and the shuttle consists of both the upper and lower layers that are mechanically connected by the silicon oxide layer, but electrically isolated. When the thermal actuator is actuated, it makes contact only with the upper layer of the shuttle. However, the lower layer of the shuttle is also moved, since both layers of the shuttle are mechanically connected through the buried oxide, actuating the scissor mechanism that is defined on the lower layer. Thus, the connection of the tuning unit to the torsional shafts does not cause any charge leakage from the electrostatic actuator, since the only connection between the electrostatic resonator side and the thermal actuator side is through the buried oxide layer between the upper and lower layers of the shuttle.

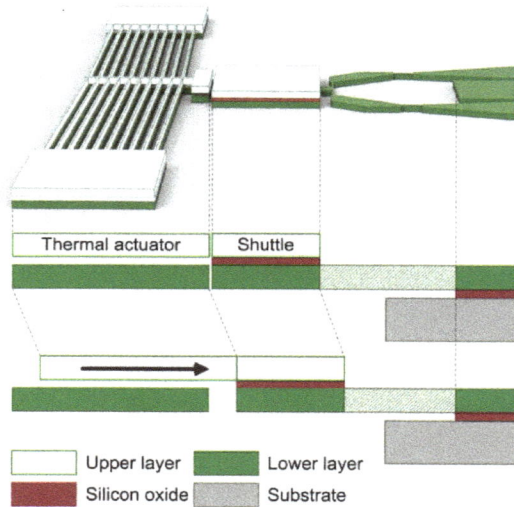

Figure 4. Electrical isolation design to avoid short circuit formation and charge leakage between the chevron thermal actuator and the electrostatic actuator.

3. Finite Element Analysis

For the design of the thermal chevron actuator, which may generate a limited amount of force, the stiffness of the scissor mechanism should be taken into consideration. Hence the finite element analysis (FEA) is conducted to calculate the required force from the thermal chevron actuator and to decide the proper dimensions of the scissor mechanism. According to the FEA results, the shaft-holder and torsional shaft start to contact with the thermal actuator stroke of 3 μm and make full contact at the stroke of 7 μm (Figure 5a) under the structural dimensions shown in Table 1. The required forces from the chevron thermal actuator for initial contact and full contact are 462 μN and 2286 μN, respectively (Figure 5a). Based on this finding, the chevron actuator is designed with the structural dimensions shown in Table 1.

In addition, the stiffness changes of the torsional shafts by the both tuning modes are quantified by FEA. First, modal analysis is performed for the shaft-widening type with the structural dimensions shown in Table 1. Based on the FEA result, we calculate torsional stiffness changes as a function of the displacement of the shuttle (i.e., stroke of the thermal actuator). As the shuttle is pulled by the thermal actuator, the torsional stiffness of the shaft gradually increases, as shown in Figure 5b. When the shuttle is moved by 7 μm, the torsional stiffness of the shaft increases by 14.0%, which corresponds to the tuning ratio of 3.4%.

For analysis for the shaft-holding type, we supposed two different restriction modes that may possibly occur when the shaft-holding flexure makes contact with the torsion bar. The schematics in Figure 6 are the cross-sectional view of the torsional shaft and the shaft-holding flexure, showing the two restriction modes. In the first mode (Figure 6a), it is assumed that the shaft-holding flexure is torsionally deformed together with the torsion bar as if there is no slip or separation between the two structures (hereafter referred to as 'no separation mode'). The FEA results for this boundary condition give a maximum stiffness change of 333.6% and a maximum resonant frequency increase of 63.3% compared to the original unrestricted structures (Figure 5b). The second mode, as depicted in Figure 6b, allows mechanical separation between the torsional shaft and the shaft-holding flexure, such that the mechanical restriction force is applied to the torsional shaft in the opposite direction of the shaft deformation, but the shaft-holding flexure is not torsionally deformed (hereafter referred to as 'separation mode'). In this case, the maximum stiffness change is 31.6% and the maximum resonant frequency increase is 7.6%, showing the reduced restriction effect on the torsional shaft (Figure 5c).

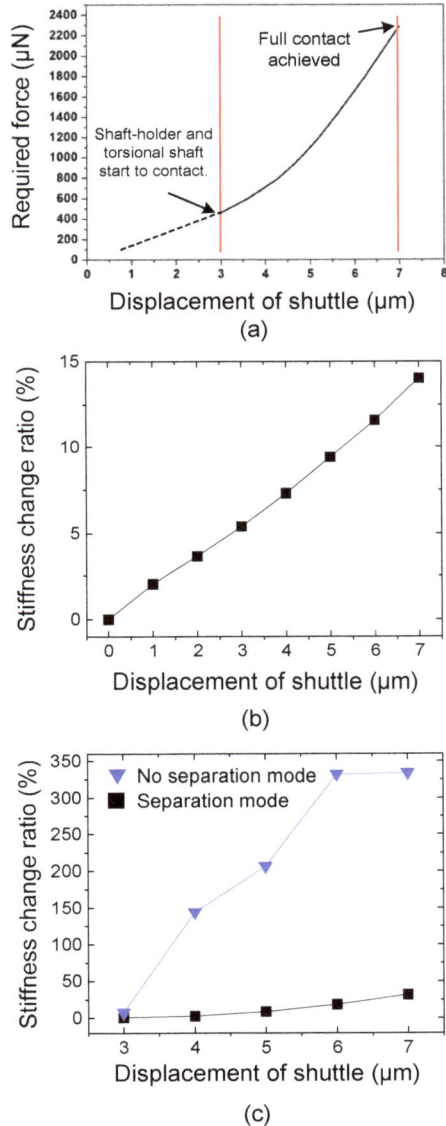

Figure 5. Finite element analysis (FEA) results: (**a**) Required force from the chevron thermal actuator to achieve full-contact between the shaft-holder and torsional shaft (shaft-holding type); (**b**) Estimated torsional stiffness change by shaft-widening type (**c**) Estimated torsional stiffness change by shaft-holding type. 'No separation mode' and 'separation mode' correspond to the restriction modes illustrated in Figure 6a,b, respectively.

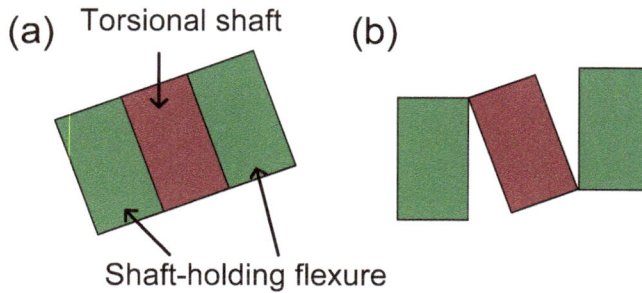

Figure 6. The cross-sectional view of the torsional shaft and shaft-holding flexure, showing two different restriction modes between the torsion bar and shaft-holding flexure. (**a**) No separation mode: the shaft-holding flexure is torsionally deformed together with the torsional shaft. The areal contact is maintained; (**b**) Separation mode: the areal contact is not maintained and the separation between the torsional shaft and shaft-holding flexure exist.

Table 1. Design parameters for the torsional resonator and tuning unit.

	Mirror diameter	800 µm
	Length/width of straight shaft	1050 µm /10 µm
	Length/width of tilted shaft	600 µm /10 µm
Torsional Resonator	Length/width of Comb	200 µm /5 µm
	The number of moving/fixed Comb	72/73
	Thickness	20 µm
	Total length/width	2240 µm /680 µm
Scissor Mechanism	Length/width of shuttle	300 µm /300 µm
	Length/width of hinges	60 µm /5 µm
Shaft-Holder	Total length	1000 µm
	Width of shaft-holding flexure	5 µm
	Length/width of chevron beam	700 µm /10 µm
Chevron Thermal Actuator	Number of chevron beam	20
	Angle of chevron beam	1.06°

4. Fabrication

Figure 7 illustrates the fabrication process for the torsional resonators and frequency tuning units [26]. Both of the structures are fabricated on DSOI wafer purchased from Ultrasil Corporation (Hayward, CA, USA). This wafer has two 20 µm-thick silicon device layers on the 425 µm thick base substrate, each of which is separated by buried oxide. The arsenic-doped silicon device layer has a resistivity less than 0.005 $\Omega \cdot$cm. To be used as etch masks, 1 µm-thick silicon oxide layers are grown on both the front- and backside of the wafer by wet oxidation process (Figure 7a). Additionally, a 2-µm-thick silicon oxide layer is deposited on the backside by plasma enhanced chemical vapor deposition (PECVD) to get a thicker backside etch mask (Figure 7b). Then the silicon oxide layers on both sides are lithographically patterned (Figure 7c). Next, the photoresist layer is patterned as shown in Figure 7d, which will be transferred to the lower device layer later. Between step (c,d), a rough alignment is acceptable, since the following oxide etch will form self-alignment between the silicon oxide and the photoresist layers (Figure 7e). After both of the device layers and the oxide layers are etched by deep reactive ion etching and reactive ion etching, respectively (Figure 7f–j), backside holes are defined (Figure 7k). Finally, to release the device, the remaining oxide etch masks and the buried oxide layers are wet-etched by exposing the entire chip to hydrofluoric acid (HF) solution (Figure 7l).

Silicon layer   Silicon oxide layer   Photoresist

(a) Wet oxidation for silicon oxide etch masks on both sides

(b) PECVD for thicker oxide layer on backside

(c) Patterning of silicon oxide etch mask

(d) Patterning of PR etch mask

(e) Self alignment by RIE

(f) Etching of upper device layer by DRIE

(g) Etching of silicon oxide etch mask by RIE

(h) Etching of lower device layer by DRIE

(i) Removal of PR etch mask

(j) Additional etching of upper device layer by DRIE

(k) Etching of handle layer by DRIE

(l) Oxide wet etching to release the movable structures

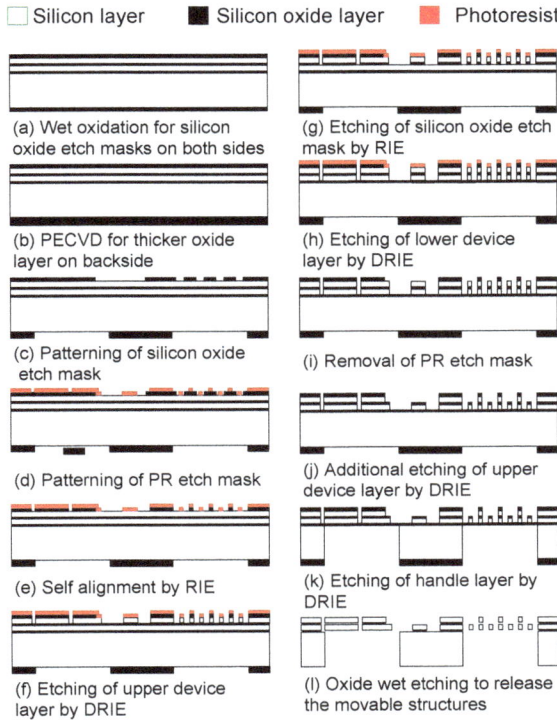

Figure 7. Fabrication process for the torsional resonator integrated with frequency tuning unit. The device is fabricated on a double silicon-on-insulator (DSOI) wafer. Reproduced with permission from [26].

5. Result

Figure 8a,b show scanning electron microscope (SEM) images of the fabricated devices for the shaft-widening type and the shaft-holding type [26]. The darker structures in the image are on the lower silicon layer, and the brighter structures are on the upper layer. The resonant frequency tuning of the electrostatic torsional resonator is experimentally verified while the driving voltages of 5 V_{ac} and 10 V_{dc} are applied on the resonator. To measure the rotational angle of the resonator, we focused a laser beam emitted from a laser diode on the micromirror of the resonator. The reflected laser beam scanned across a certain distance on a target screen. We measured the scanning distance of the laser beam (R), and calculated the rotational angle (θ) of the resonator as follows:

$$\theta = \tan^{-1}\left(\frac{R}{2D}\right)$$

where D is the distance between the target screen and the mirror.

Figure 9 shows how the frequency response of the resonator is changed from the untuned state to differently-tuned states by the shaft-widening type. The frequency response of our system does not perfectly comply with a form of Lorentzian curve; the stiffness-softening behavior is observed in all the frequency spectra, and is represented by a slight declination of the resonant peak toward the low frequency region. Due to this asymmetry, the left and right sides of the resonant frequency are fitted separately. We suspect that this nonlinearity mainly originated from the electrostatic actuation in our system [14]. However, obvious bifurcation or jump discontinuity due to the nonlinearity is not observed in the frequency spectra, thus we expect that the developed system will function well even in the presence of nonlinearity, as long as the excitation force no longer increases from the applied value. When the tuning DC voltages of 10 V and 12 V are applied to the shaft-widening type, the resonant frequency is shifted to 1.560 kHz and 1.593 kHz, respectively, from the untuned resonant frequency of 1.507 kHz (Table 2). The tuning ratios are 3.31% and 5.29% at tuning voltages of 10 V and 12 V, respectively.

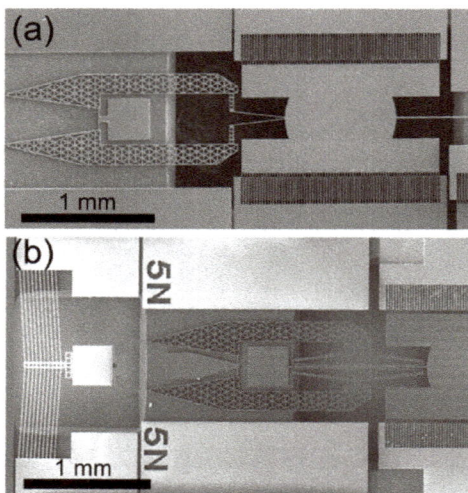

Figure 8. Scanning electron microscope (SEM) images of the fabricated devices: (**a**) Shaft-widening type and (**b**) shaft-holding type. Reproduced with permission from [26].

In the shaft-holding type, when the tuning voltage of 7 V is applied to the thermal actuator, the shaft-holding flexures started to make contact with the torsional spring, and under 12 V, they formed full contact. Between 7 V and 12 V, the mechanical restriction process is completely reversible, and continuous frequency tuning up and down is achieved. The experimental measurement results for resonant frequency tuning are presented in Table 3, and changes in frequency responses are plotted in Figure 10 [26]. Under the tuning DC voltage of 8 V, 10 V and 12 V, the resonant frequency is shifted from the untuned resonant frequency of 1.698 kHz to 1.749 kHz, 1.826 kHz and 1.880 kHz, respectively, which correspond to the tuning ratio of 3.03%, 7.05% and 10.7%, respectively. The untuned resonant frequency of the torsional resonator (1.698 kHz) exhibits a deviation of 9.25% from the designed resonant frequency (1.871 kHz). The maximum tuning ratio of the shaft-holding type (10.7%) is sufficient to compensate for this discrepancy. However, it is noteworthy to mention that geometric error could be reduced under well-controlled fabrication conditions. The manufacturing processes for commercialized resonators probably have less fabrication errors than ours. In fact, as we mentioned earlier, it has been reported that the resonant frequencies of micromachined resonators generally exhibit a deviation between $\pm1\%\sim\pm5\%$ [7]. The Microfabrication Laboratory at UC-Berkeley have also reported a maximum deviation of 4.9% in resonant frequencies for batch-fabricated resonators [6]. Based on these criteria, the shaft-holding type, which exhibits lower tuning performance (maximum tuning ratio of 5.29%) than the shaft-holding type, also meets the needs of demand. It is expected that the tuning ratio would increase for the both types if the tuning actuator could be further operated, however, it could not be experimentally verified since the chevron thermal actuator burnt at a tuning voltage greater than 12 V.

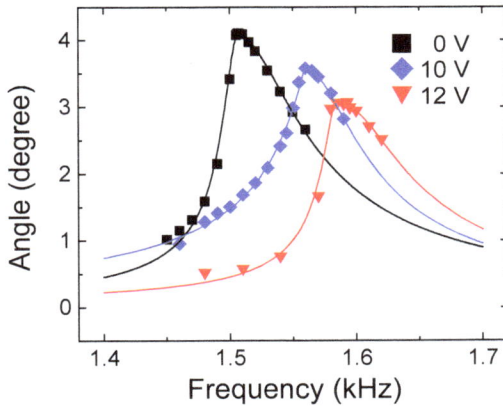

Figure 9. Frequency response change achieved by the shaft-widening type. Data points are fitted with the Lorentzian function.

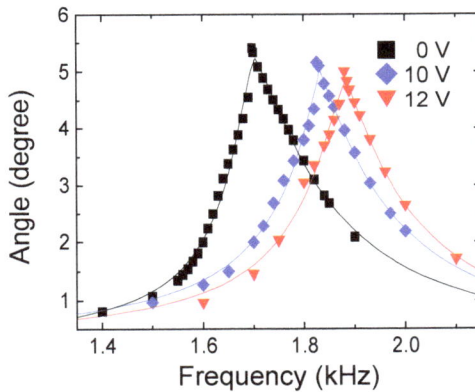

Figure 10. Frequency response change achieved by the shaft-holding type. Data points are fitted with the Lorentzian function. Reproduced with permission from [26].

Table 2. Resonant frequency shifts achieved by the shaft-widening type.

Tuning Voltage (V)	Resonant Frequency (kHz)	Tuning Ratio (%)
0 V	1.507	-
10 V	1.560	3.31
12 V	1.593	5.29

Table 3. Resonant frequency shifts achieved by the shaft-holding type. Reproduced with permission from [26].

Tuning Voltage (V)	Resonant Frequency (kHz)	Tuning Ratio (%)
0 V	1.698	-
8 V	1.749	3.03
10 V	1.826	7.05
12 V	1.880	10.7

Figure 11 compares measured and simulated resonant frequency changes by the frequency tuning units. Both the measured and the simulated resonant frequencies increase with the application of tuning voltage, but there is some discrepancy in values between them. For example, the simulated values for the shaft-widening type differ from the experimental results up to 34.8% (Figure 11a). Our study does not further verify the factors causing this discrepancy. However, it is worth exploring possible sources. Dimensional error from the microfabrication process could be one of the sources of this mismatch. Imperfect lithography and the ion-reactive process might change the performance of the tuning unit, if the resonant frequency is shifted from the desired value. It is also worth noting that we used simplified boundary conditions for FEA. For example, when we derive the resonant frequency by modal analysis, we do not consider some conditions that may cause nonlinearity, including relatively large deflection, sliding between contact surfaces, varying contact area, and the electrostatic spring effect.

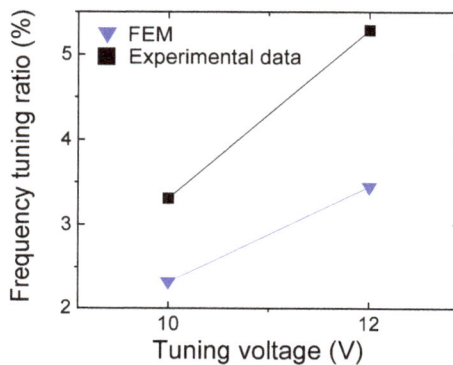

(a) Shaft-widening type

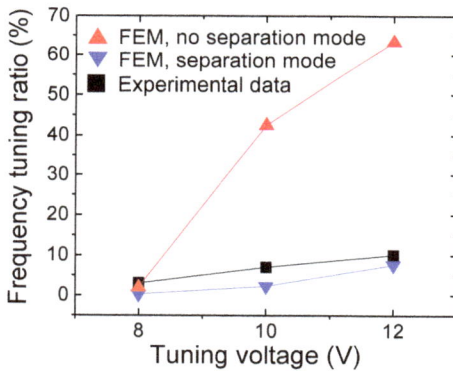

(b) Shaft-holding type

Figure 11. Measured and simulated results for resonant frequency change as a function of applied tuning voltage: (**a**) The shaft-widening type; (**b**) the shaft-holding type.

In the previous section, we assumed two restriction modes in the shaft-holding type: no separation and separation modes. At a tuning voltage of 12 V, the estimated frequency tuning ratios for the no separation and separation mode are 63.3% and 7.6%, respectively (Figure 11b). It is not obvious which

Micromachines **2017**, *8*, 342

mode is really happening, and a co-existence of both modes is also possible. However, the frequency tuning ratio of the no separation mode is closer to that of the experimental results (10.7% at 12 V).

Since our tuning mechanism is based on mechanical stress and contact, we have extracted Q-factors for each frequency spectrum. Instead of a half-power bandwidth formula, which is generally used to calculate Q-factors for linear oscillating system, we used an alternative formula that is modified for a non-linear system, especially for a micro-scanning mirror [29]. The Q-factors of the shaft-widening type are calculated as 28, 27, and 29 at applied tuning voltages of 0 V, 10 V, and 12 V, respectively. Considering that the Q-factor remains almost constant as the tuning is applied, it seems that the level of energy dissipation does not change significantly with the increment of the stress inside the spring. The Q-factors of the shaft-holding type are 17, 16, and 16 at applied tuning voltages of 0 V, 10 V, and 12 V, respectively, showing a slight decrease even under mechanical restriction. One of the possible scenarios explaining this is that the resonator is under the dominant influence of the air damping, since a relatively large mass (i.e., the micromirror) is oscillating in the ambient environment. Thus, the contribution of mechanical stress or contact to the overall energy dissipation may not be notable.

6. Conclusions

In this work, the resonant frequency tuning of a torsional resonator has been demonstrated by two concepts: shaft-widening and shaft-holding. Based on the FEA results, we verified the validity of our tuning methods, and derived the estimated change of torsional stiffness and tuning range. The frequency tuning units were integrated into the resonator, and there is no necessity for additional post fabrication processes for frequency-tuning. We experimentally verified the performance of both designs. The results indicate that the shaft-holding type has a wider tuning range compared to the shaft-widening type. Nevertheless, the tuning ratios of both are higher compared to those of the previously method [14], and are also sufficient to compensate for the dimensional errors of the conventional microfabrication process [6,7].

Acknowledgments: This research was supported by the Korea Institute of Energy Technology Evaluation and Planning (KETEP) and the Ministry of Trade, Industry & Energy (MOTIE) of the Republic of Korea (No. 20163030024420), and the Basic Science Research Program through the National Research Foundation of Korea (NRF) funded by the Ministry of Science, ICT and Future Planning (NRF-2015R1A2A1A01005496).

Author Contributions: Jongbaeg Kim and Jae-Ik Lee conceived and designed the experiments; Jae-Ik Lee and Sunwoo Park performed the experiments. Jae-Ik Lee, Youngkee Eun, and Bongwon Jeong analyzed the data; Jongbaeg Kim and Jae-Ik Lee wrote the paper.

Conflicts of Interest: The authors declare no conflict of interest.

References

1. Zhang, W.M.; Hu, K.M.; Peng, Z.K.; Meng, G. Tunable micro- and nanomechanical resonators. *Sensors* **2015**, *15*, 26478–26566. [CrossRef] [PubMed]
2. Toshiyoshi, H.; Fujita, H. Electrostatic micro torsion mirrors for an optical switch matrix. *J. Microelectromech. Syst.* **1996**, *5*, 231–237. [CrossRef]
3. Hwang, K.S.; Lee, S.M.; Kim, S.K.; Lee, J.H.; Kim, T.S. Micro- and nanocantilever devices and systems for biomolecule detection. *Annu. Rev. Anal. Chem.* **2009**, *2*, 77–98. [CrossRef] [PubMed]
4. Madou, M.J. *Fundamentals of Microfabrication and Nanotechnology*, 3rd ed.; CRC Press: Boca Raton, FL, USA, 2012; ISBN 978-08-49331-80-0.
5. Tang, W.C.; Nguyen, T.-C.H.; Judy, M.W.; Howe, R.T. Electrostatic-comb drive of lateral polysilicon resonators. *Sens. Actuators A Phys.* **1990**, *21*, 328–331. [CrossRef]
6. Chiao, M.; Lin, L. Post-packaging frequency tuning of microresonators by pulsed laser deposition. *J. Micromech. Microeng.* **2004**, *14*, 1742–1747. [CrossRef]

7. Hsu, W.-T.; Brown, A.R. Frequency trimming for mems resonator oscillators. In Proceedings of the IEEE International Conference on Frequency Control Symposium, Geneva, Switzerland, 29 May–1 June 2007; pp. 1088–1091.

8. Yong, Y.-K.; Vig, J.R. Resonator surface contamination—A cause of frequency fluctuations? *IEEE Trans. Ultrason. Ferroelectr. Freq. Control* **1989**, *36*, 452–458. [CrossRef] [PubMed]

9. Koskenvuori, M.; Mattila, T.; Häärä, A.; Kiihamäki, J.; Tittonen, I.; Oja, A.; Seppä, H. Long-term stability of single-crystal silicon microresonators. *Sens. Actuators A Phys.* **2004**, *115*, 23–27. [CrossRef]

10. Enderling, S.; Hedley, J.; Jiang, L.D.; Cheung, R.; Zorman, C.; Mehregany, M.; Walton, A.J. Characterization of frequency tuning using focused ion beam platinum deposition. *J. Micromech. Microeng.* **2006**, *17*, 213–219. [CrossRef]

11. Joachim, D.; Lin, L. Characterization of selective polysilicon deposition for mems resonator tuning. *J. Microelectromech. Syst.* **2003**, *12*, 193–200. [CrossRef]

12. Syms, R.R.A.; Moore, D.F. Focused ion beam tuning of in-plane vibrating micromechanical resonators. *Electron. Lett.* **1999**, *35*, 1277–1278. [CrossRef]

13. Witkamp, B.; Poot, M.; Pathangi, H.; Hüttel, A.; Van der Zant, H. Self-detecting gate-tunable nanotube paddle resonators. *Appl. Phys. Lett.* **2008**, *93*, 111909. [CrossRef]

14. Eun, Y.; Kim, J.; Lin, L. Resonant-frequency tuning of angular vertical comb-driven microscanner. *Micro Nano Syst. Lett.* **2014**, *2*, 4. [CrossRef]

15. Lee, W.-S.; Kwon, K.-C.; Kim, B.-K.; Cho, J.-H.; Youn, S.-K. Frequency-shifting analysis of electrostatically tunable micro-mechanical actuator. *CMES Comp. Model. Eng. Sci.* **2004**, *5*, 279–286.

16. Gallacher, B.J.; Hedley, J.; Burdess, J.S.; Harris, A.J.; Rickard, A.; King, D.O. Electrostatic correction of structural imperfections present in a microring gyroscope. *J. Microelectromech. Syst.* **2005**, *14*, 221–234. [CrossRef]

17. Lee, K.B.; Cho, Y.-H. A triangular electrostatic comb array for micromechanical resonant frequency tuning. *Sens. Actuators A Phys.* **1998**, *70*, 112–117. [CrossRef]

18. Lee, K.B.; Lin, L.; Cho, Y.-H. A closed-form approach for frequency tunable comb resonators with curved finger contour. *Sens. Actuators A Phys.* **2008**, *141*, 523–529. [CrossRef]

19. Jensen, B.D.; Mutlu, S.; Miller, S.; Kurabayashi, K.; Allen, J.J. Shaped comb fingers for tailored electromechanical restoring force. *J. Microelectromech. Syst.* **2003**, *12*, 373–383. [CrossRef]

20. Morgan, B.; Ghodssi, R. Vertically-shaped tunable mems resonators. *J. Microelectromech. Syst.* **2008**, *17*, 85–92. [CrossRef]

21. Scheibner, D.; Mehner, J.; Reuter, D.; Kotarsky, U.; Gessner, T.; Dötzel, W. Characterization and self-test of electrostatically tunable resonators for frequency selective vibration measurements. *Sens. Actuators A Phys.* **2004**, *111*, 93–99. [CrossRef]

22. Remtema, T.; Lin, L. Active frequency tuning for micro resonators by localized thermal stressing effects. *Sens. Actuators A Phys.* **2001**, *91*, 326–332. [CrossRef]

23. Syms, R.R. Electrothermal frequency tuning of folded and coupled vibrating micromechanical resonators. *J. Microelectromech. Syst.* **1998**, *7*, 164–171. [CrossRef]

24. Elata, D.; Leus, V.; Hirshberg, A.; Salomon, O.; Naftali, M. A novel tilting micromirror with a triangular waveform resonance response and an adjustable resonance frequency for raster scanning applications. In Proceedings of the TRANSDUCERS Solid-State Sensors, Actuators and Microsystems Conference, Lyon, France, 10–14 June 2007; pp. 1509–1512.

25. Shmilovich, T.; Krylov, S. Linear tuning of the resonant frequency in tilting oscillators by an axially loaded suspension flexure. In Proceedings of the IEEE 21st International Conference on Micro Electro Mechanical Systems, Wuhan, China, 13–17 January 2008; pp. 657–660.

26. Lee, J.-I.; Park, S.; Eun, Y.; Jeong, B.; Kim, J. Resonant frequency tuning of torsional microscanner by mechanical restriction using mems actuator. In Proceedings of the IEEE 22nd International Conference on Micro Electro Mechanical Systems, Sorrento, Italy, 25–29 January 2009; pp. 164–167.

27. Ya'akobovitz, A.; Krylov, S.; Shacham-Diamand, Y. Large angle SOI tilting actuator with integrated motion transformer and amplifier. *Sens. Actuators A Phys.* **2008**, *148*, 422–436. [CrossRef]

28. Lott, C.D.; McLain, T.W.; Harb, J.N.; Howell, L.L. Modeling the thermal behavior of a surface-micromachined linear-displacement thermomechanical microactuator. *Sens. Actuators A Phys.* **2002**, *101*, 239–250. [CrossRef]

29. Davis, W.O. Measuring quality factor from a nonlinear frequency response with jump discontinuities. *J. Microelectromech. Syst.* **2011**, *20*, 968–975. [CrossRef]

micromachines

MDPI

Article

Static and Dynamic Mechanical Behaviors of Electrostatic MEMS Resonator with Surface Processing Error

Jingjing Feng [1,2,3,*], Cheng Liu [1,3], Wei Zhang [2,*] and Shuying Hao [1,3,*]

1   Tianjin Key Laboratory for Advanced Mechatronic System Design and Intelligent Control,
    School of Mechanical Engineering, Tianjin University of Technology, Tianjin 300384, China;
    163110303@stud.tjut.edu.cn
2   Beijing Key Laboratory on Nonlinear Vibrations and Strength of Mechanical Structures,
    Beijing University of Technology, College of Mechanical Engineering, Beijing 100124, China
3   National Demonstration Center for Experimental Mechanical and Electrical Engineering Education,
    Tianjin University of Technology, Tianjin 300384, China
*   Correspondence: jjfeng@tju.edu.cn (J.F.); wzhang@bjut.edu.cn (W.Z.); syhao@tju.edu.cn (S.H.);
    Tel.: +86-226-021-4133 (J.F. & S.H.); +86-106-739-2867 (W.Z.)

Received: 30 November 2017; Accepted: 12 January 2018; Published: 17 January 2018

Abstract: The micro-electro-mechanical system (MEMS) resonator developed based on surface processing technology usually changes the section shape either due to excessive etching or insufficient etching. In this paper, a section parameter is proposed to describe the microbeam changes in the upper and lower sections. The effect of section change on the mechanical properties is studied analytically and verified through numerical and finite element solutions. A doubly-clamped microbeam-based resonator, which is actuated by an electrode on one side, is investigated. The higher-order model is derived without neglecting the effects of neutral plane stretching and electrostatic nonlinearity. Further, the Galerkin method and Newton–Cotes method are used to reduce the complexity and order of the derived model. First of all, the influence of microbeam shape and gap variation on the static pull-in are studied. Then, the dynamic analysis of the system is investigated. The method of multiple scales (MMS) is applied to determine the response of the system for small amplitude vibrations. The relationship between the microbeam shape and the frequency response is discussed. Results show that the change of section and gap distance can make the vibration soften, harden, and so on. Furthermore, when the amplitude of vibration is large, the frequency response softening effect is weakened by the MMS. If the nonlinearity shows hardening-type behavior at the beginning, with the increase of the amplitude, the frequency response will shift from hardening to softening behavior. The large amplitude in-well motions are studied to investigate the transitions between hardening and softening behaviors. Finally, the finite element analysis using COMSOL software (COMSOL Inc., Stockholm, Sweden) is carried out to verify the theoretical results, and the two results are very close to each other in the stable region.

Keywords: MEMS; processing error; section parameter; nonlinear vibration

---

1. Introduction

Electrostatically-actuated microbeams have become major components in many micro-electro-mechanical system (MEMS) devices [1] such as switches [2,3], sensors [4,5] and resonators [6] due to their geometric simplicity, broad applicability and easy to implement characteristics. Moreover, the existence of structure nonlinearity and nonlinear electrostatic force can make microbeams exhibit rich static and dynamic behaviors [7,8]. These behaviors have aroused the interest of many scholars, who have joined the study of MEMS. However, most of them have been aiming at the equal section beam under ideal conditions. However, the microbeams and microdiaphragms fabricated through surface processing technology are prone to errors during fabrication [9]. Such errors during fabrication of microdevices cannot be ignored as they can cause the bending of the microbeam neutral surface and change the width, thickness and gap distance of the microresonator. Hence, it is very important to analyze the static and dynamic behaviors of the electrostatically-actuated beam with surface processing error for understanding its global dynamic behavior, developing a dynamic control problem and optimizing vibration design. In the present paper, a doubly-clamped beam of variable thickness actuated by a one-sided electrode is considered to study the influence of section variation on static and dynamic behaviors.

The static pull-in instability is one of the key issues in the design of MEMS [10,11]. When the direct current (DC) voltage is increased beyond a critical value, the stable equilibrium positions of the microbeam cease to exist, and the pull-in instability will be triggered [12,13]. For example, Abdel-Rahman et al. [14] investigated an electrically-actuated microbeam accounting for midplane stretching and derived the static pull-in position of microbeam. Younis et al. [15] studied the effect of residual stress on the static pull-in of microresonators and found that the residual stress would increase the pull-in voltage. The system's vibrations can cause an interesting nonlinear phenomenon such as hysteresis, softening behavior, snap through and dynamic pull-in instability when it is excited with DC and alternating current (AC) voltages [16–18]. These analyses are helpful to further grasp the dynamic instability of microcomponents. Ghayesh and Farokh [19] investigated the static and dynamic behavior of an electrically-actuated MEMS resonator based on the modified couple stress theory. It is found that the pull-in voltage is larger by the coupled correction theory compared with the classical theory. Zhang et al. [20] used the method of multiple scales (MMS) to study the response and dynamic behaviors of the resonant parameters resonant in the MEMS resonator. The softening behavior of the DC voltage and the effect of damping on the frequency response curve were discussed. Ibrahim [21] investigated the effect of nonlinearities of a capacitive accelerometer due to squeeze film damping (SQFD) and electrostatic actuation by the theoretical and experimental methods. Theoretical results are compared to experimental data showing excellent agreement. Ghayesh, Farokh and Gholipour [22,23] investigated the nonlinear dynamics of a microplate based on the modified couple stress theory. The influence of system parameters on the resonant responses was highlighted by the frequency-response and force-response curves. Alsaleem et al. [24] conducted an experimental study to understand the dynamic pull-in voltage of the electrostatic drive microresonator. The experimental and theoretical results are in good agreement. Furthermore, the applicable microresonator conditions are pointed out. However, the majority of the previous studies are based on the equal section microbeam, i.e., the impact of model errors is neglected.

With the deepening research in this discipline, the importance of understanding the processing error of the microbeam on its performance has been realized. There are several sources of errors possible, for example, residual [25], initial offset imperfection [26], surface processing technology precision [27], etc. The residual stress causes the bending of the microbeam to form a microarch [28]. Farokhi and Ghayesh [29,30] established the mathematical model of a geometrically imperfect microbeam/microplate, the nonlinear force of which was actuated on the basis of the modified couple stress theory. The influence of physical parameters on the natural frequency and frequency response were analyzed. What is more, Farokhi and Ghayesh [31] investigated the three-dimensional motion characteristics of perfect and imperfect Timoshenko microbeams under mechanical and thermal forces. Ruzziconi [32] studied many kinds of nonlinear behaviors of the microarch under different parameters through theoretical analysis. The conclusions are in good agreement with experimental analysis. Besides, a reliable theoretical model was obtained. Ruzziconi [33] predicted the global bifurcation of the electrostatically-actuated microarch and studied the complex dynamics of the microarch. Krylov et al. [34] studied the electrostatically-actuated microarch structures, focusing on the influence of system geometry parameters on its dynamic behaviors and found that the main influencing factors are microarch thickness, microarch height and the distance between the microarch and the plate. Xu et al. [35] studied the dynamic behavior of clamped-clamped carbon nanotubes with initial bending and explored the non-resonance and resonance of carbon nanotubes by the shooting method. Hassen et al. [36] established a clamped-clamped beam model considering the initial bending and obtained its static and dynamic response using the Galerkin method and MMS. The microbeam with initial offset imperfections is usually actuated by two electrodes. In this case, the microbeam is still rectangular. However, the initial offset imperfections can break the symmetry along the transverse vibrational direction in dynamic MEMS devices. Mobki et al. [37] discussed the influence of the initial offset imperfections on the static bifurcation of a MEMS resonator. Han et al. [38] considered the effect of initial offset imperfections on the mechanical behaviors of microbeam. The global static and dynamic analysis of the microresonator is carried out using MMS and the finite difference method. Results show that the initial offset may induce a complex frequency rebound phenomenon, and there exists the frequency response in the medium and large amplitude in-well transitions between softening and hardening behaviors. Although these two kinds of error forms have a great influence on the mechanical properties of the system, the microbeam model is still an equal section beam. The error caused due to the accuracy of surface machining will change the width or thickness of the microbeam. Such dimensional changes affect the structural stiffness and electrostatic force, so it is necessary to study these. However, the influence of such errors on the shape of the microbeam is random. Therefore, scholars usually do smoothing processing by setting up the parametric equation model and adjusting the shape of the microbeam by changing the parameters. Herrera [39] studied the resonant behavior of a single-layered variable section microbeam. Furthermore, scholars have attempted to optimize the MEMS device by optimizing the equation parameters. Joglekar and Trivedi [40,41] proposed a versatile parametric width function, which can smoothly vary the width of a clamped-clamped microbeam along its length. The parameters of the width function are optimized, and the methodology is demonstrated in several cases [40,41]. On this basis, Zhang [42] discussed the effects of the optimized shape on the dynamic response of the microbeam. Few researchers have considered the influence of variations in microbeam thickness on the mechanical behavior. Kuang and Chen [43] investigated the effect of shaping the thickness of a microactuator and gap distance on its natural frequencies. Their study concluded that the shape variation could significantly alter the dynamic behavior of the microbeam. In particular, the working voltage range was increased six times as compared to a uniform rectangular cross-section microbeam with a flat electrode. Najar et al. [27,44] simulated and analyzed the deflection and motion of variable section beams in MEMS devices, and the effect of changing their geometrical parameters on the static bifurcation and frequency response was observed. However, only single-sided section changes were considered in their study. The sections of the microbeam both change up and down by taking into account the actual processing result. One section change is merely applied to special cases. Therefore, in this paper, simultaneous changes in two sections of the microbeam are considered to ensure that the obtained research models will be closer to reality. In addition, only the static pull-in voltage and frequency response were studied in [27,44]. The nonlinear softening and hardening behaviors, spring softening and other non-linear behaviors are still unclear. It is also essential to study the scope of applications of the theoretical analysis. In this work, the influence of processing error on the nonlinear softening and hardening behaviors, electrostatic softening and dynamic behavior with a large amplitude are analyzed.

The structure of this paper is as follows. In Section 2, the model (partial differential equations) based on the electrostatically-driven microbeam considering the size effect is given. A parameter is proposed for describing the variations in the microbeam section to study the degree of influence. The Galerkin method and Newton–Cotes method are applied to transform the original equation into the ordinary differential equation. In Section 3, the effects of section parameters and gap distance on the static pull-in and safety area are discussed. In Section 4, the MMS is applied to determine the response of the system under small amplitude vibrations. The relationship between section parameters and nonlinear characteristics and the relationship between the section parameters and the transitions between softening and hardening are discussed. In Section 5, the results obtained using

COMSOL (COMSOL Inc., Stockholm, Sweden) simulations are presented to verify the theory. Finally, the summary and conclusions are presented in the last section.

2. Mathematical Model

2.1. Governing Equation

In this paper, a model considering the effect of surface machining error on the thickness of the microbeam is studied. The bending vibration equation of the system is obtained through force analysis.

The schematic diagram of microbeam is shown in Figure 1. The thickness of the microbeam is not constant due to the processing errors. In this study, a section parameter λ is proposed. The shape of the microbeam is controlled by adjusting the value of section parameter λ. When $\lambda < 0$, the thickness of two clamped ends is greater than the thickness of the middle portion. When $\lambda > 0$, the thickness gradually decreases from the middle to both of the clamped ends. The $\lambda = 0$ case is the ideal case where the section beam thickness is uniform.

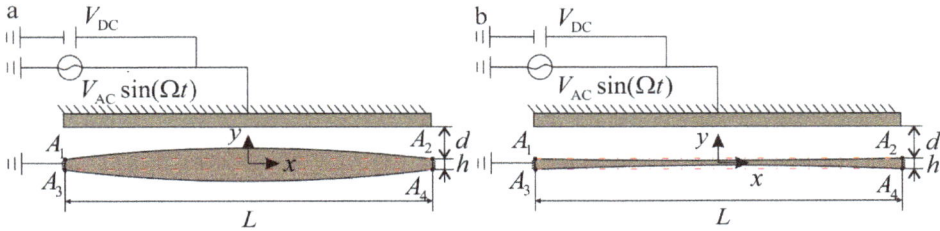

Figure 1. Schematic of an electrically-actuated microbeam. (**a**) $\lambda > 0$ case. (**b**) $\lambda < 0$ case. The dotted line in red is the ideal section location of the beam.

Since the pull-in behavior will cause structural damage, such instability should be avoided in microresonators. Stability can be ensured by considering the impact of processing errors on the pull-in effect.

The thickness of the microbeam changes according to $y_1(x) = \frac{h}{2} + \lambda h \sin \frac{\pi x}{L}$ and $y_2(x) = \{-\}(\frac{h}{2} + \lambda h \sin \frac{\pi x}{L})$ after considering the effect of surface processing error. $y_1(x)$ is a function consisting of the curve $A_1 A_2$, and $y_2(x)$ is a function consisting of the curve $A_3 A_4$. As shown in Figure 1, A_1, A_2 and A_3, A_4 represent the end points at the upper and lower sections, respectively. The parameter λ will be investigated to observe its impact on the system. The cross-sectional area $A(x) = A_0(1 + 2\lambda \sin(\frac{\pi x}{L}))$ and moment of inertia $I(x) = I_0(1 + 2\lambda \sin(\frac{\pi x}{L}))^3$ are calculated from $y_1(x)$ and $y_2(x)$. $A_0 = bh$ and $I_0 = \frac{bh^3}{12}$ are the area and the moment of inertia of the two clamped sides, respectively. L and b are the length and the width of the microbeam, respectively. h is the thickness of the microbeam at the clamped ends. c is the damping coefficient of the system. d is the distance from the board to the x-axis. E is the effective Young's modulus, and ρ is the material density. The actuation of the clamped-clamped microbeam is realized by using a bias voltage V_{DC} and AC voltage $V_{AC} \cos$ (Ωt). Ω is the alternating current excitation frequency. In the microresonator, V_{AC} is far less than V_{DC}. ε_0 is the dielectric constant in the free space, and ε_r is the relative permittivity of the gap space medium with respect to the free space.

The equation of motion that governs the transverse deflection $y(x,t)$ is written as [27]:

$$\frac{\partial^2}{\partial x^2}\left(EI(x)\frac{\partial^2 y}{\partial x^2}\right) + \rho A(x)\frac{\partial^2 y}{\partial t^2} + c\frac{\partial y}{\partial t} = \frac{E}{2L}\int_0^1 A(x)\left(\frac{\partial y}{\partial x}\right)^2 dx \frac{\partial^2 y}{\partial x^2} + \frac{\varepsilon_0 \varepsilon_r b[V_{DC} + V_{AC}\cos(\Omega t)]^2}{2(d - y_1(x) - y)^2} \tag{1}$$

The first term on the right-hand side of Equation (1) represents the mid-plane stretching effects, and the second term represents the electrostatic force. The following are the boundary conditions:

$$y(0,t) = \frac{\partial y(0,t)}{\partial x} = 0, \quad y(L,t) = \frac{\partial y(L,t)}{\partial x} = 0 \tag{2}$$

The range of the parameter $6(d/h)^2$ is around $6(d/h)^2 \in [0.1, 10]$ in the equal cross-section microbeam resonator [1]. It can be deduced from $6(d/h)^2 \in [0.1, 10]$ that the ratio of the gap distance to the microbeam thickness ranges from 0.13–1.3 in equal cross-section microbeam resonator. After adding the parameter λ, the thickness and the gap distance become $h + 2\lambda h$ and $d - \lambda h$, respectively. Therefore, the ratio becomes $0.13 \leq \frac{d - \lambda h}{h + 2\lambda h} \leq 1.3$. Besides, the range of λ should satisfy the physical model. When the upper and lower sections of the microbeam are becoming thinner, the change of section λh cannot exceed the microbeam's neutral surface, which is $\lambda h > -\frac{h}{2}$. The change of section cannot contact the plate when the microbeam section is becoming thicker, which is $\lambda h < d - \frac{h}{2}$. Therefore, the ranges are as follows:

$$\begin{cases} 0.13 \leq \frac{d-\lambda h}{h+2\lambda h} \leq 1.3 \\ -\frac{h}{2} < \lambda h < d - \frac{h}{2} \end{cases} \tag{3}$$

It can be understood from Equation (3) that when $\lambda = 0$, the range of d/h is $0.12 \leq d/h \leq 1.3$. By simplification:

$$-0.3 \leq 0.27\frac{d}{h} - 0.37 \leq \lambda \leq 0.79\frac{d}{h} - 0.10 \leq 0.9 \tag{4}$$

For convenience, the following non-dimensional quantities are defined:

$$\widehat{x} = \frac{x}{L}, \widehat{b} = \frac{b}{d}, \widehat{y} = \frac{y}{d}, \widehat{y}_1 = \frac{y_1(x)}{d}, \widehat{y}_2 = \frac{y_2(x)}{d}, \widehat{A}(\widehat{x}) = \frac{A(x)}{A_0}, \widehat{I}(\widehat{x}) = \frac{I(x)}{I_0}, \widehat{t} = \frac{t}{T}, \omega = \frac{\Omega t}{\widehat{t}},$$
$$T = \sqrt{\frac{l^4 \rho A_0}{E I_0}}, \mu = \frac{cL^4}{E I_0 T}, \alpha_1 = \frac{\varepsilon_0 \varepsilon_r b l^4 V_{DC}^2}{2E I_0 d^3}, \rho = \frac{V_{AC}}{V_{DC}}, \alpha_2 = 6\left(\frac{d}{h}\right)^2 \tag{5}$$

Substituting Equation (5) into Equations (1) and (2), the following non-dimensional equation of motion can be obtained:

$$\frac{\partial^2}{\partial \widehat{x}^2}\left(I(\widehat{x})\frac{\partial^2 \widehat{y}}{\partial \widehat{x}^2}\right) + A(\widehat{x})\frac{\partial^2 \widehat{y}}{\partial \widehat{t}^2} + \mu\frac{\partial \widehat{y}}{\partial \widehat{t}} - \alpha_2 \int_0^1 A(\widehat{x})\left(\frac{\partial \widehat{y}}{\partial \widehat{x}}\right)^2 dx \frac{\partial^2 \widehat{y}}{\partial \widehat{x}^2} = \frac{\alpha_1}{\left(1 - \widehat{y}_1(x) - \widehat{y}\right)^2} \tag{6}$$

with boundary conditions:

$$\widehat{y}(0, \widehat{t}) = \frac{\partial \widehat{y}(0, \widehat{t})}{\partial \widehat{x}} = 0, \widehat{y}(1, \widehat{t}) = \frac{\partial \widehat{y}(1, \widehat{t})}{\partial \widehat{x}} = 0 \tag{7}$$

In the following simplifications, the "ˆ" notation is dropped for convenience.

2.2. Galerkin Expansion

The Galerkin method is applied to derive a reduced-order model, and the deflection is expressed as:

$$y(x,t) = \sum_{i=1}^{\infty} u_i(t)\phi_i(x) \tag{8}$$

The boundary conditions are as follows:

$$\phi_i(0) = \phi_i(1) = \phi_i{'}(0) = \phi_i{'}(1) = 0 \tag{9}$$

where $u_i(t)$ is the modal coordinate amplitude of the i-th mode. $\phi_i(x)$ is the i-th mode shapes of the normalized undamped linear orthonormal. For an electrostatic actuated microbeam, a single degree-of-freedom model is sufficient to capture all the key nonlinear aspects in the Galerkin approximation [25]. However, the one-mode approximation cannot capture the mode coupling effect or internal resonances. These phenomena can be predicted to obtain a reasonable result by implementing the number of modes. Nevertheless, the analysis becomes computationally expensive. Since the main objective of this paper is to explore the main resonance problem in the nonlinear dynamics problem, the first-order mode is sufficient to obtain good results. In this paper, the first-order modal vibration $y(x,t) = u(t)\phi(x)$ is assumed. Substitute Equation (8) into Equation (6). Upon multiplying by $\phi_i(x)$ and integrating, the outcome is from $x = 0$ to 1, and one can obtain the following equation:

$$\ddot{u} + \mu\dot{u} + k_1 u - \alpha_2 k_3 u^3 = \alpha_1 \int_0^1 \frac{\phi(x)}{(1-y_1(x)-\phi(x)u)^2} dx + 2\alpha_1 \rho \cos(\omega t) \int_0^1 \frac{\phi(x)}{(1-y_1(x)-\phi(x)u)^2} dx$$
$$+ \alpha_1 \rho^2 \cos^2(\omega t) \int_0^1 \frac{\phi(x)}{(1-y_1(x)-\phi(x)u)^2} dx \tag{10}$$

where $\dot{u} = du/dt$. The symbolic meanings of μ, k_1 and k_2 are discussed in Appendix A.

2.3. Newton–Cotes Method

The integral of the electrostatic force in Equation (10) is complicated; therefore, the Newton–Cotes method is applied to fit the electrostatic force.

The integral interval $[\tilde{a},\tilde{b}]$ is divided into n equal divisions. The step length is set as $\Delta h = \frac{\tilde{b}-\tilde{a}}{n}$. The node is $x_k = \tilde{a} + k\Delta h$, where $k = 0, 1, 2, \ldots, n$. The interpolation type quadrature formula is as follows:

$$\int_{\tilde{a}}^{\tilde{b}} \frac{\phi(x)}{(1 - y_1(x) - \phi(x)u)^2} dx = (\tilde{b} - \tilde{a}) \sum_{k=0}^{n} C_k^{(n)} \frac{\phi(x_k)}{(1 - y_1(x_k) - \phi(x_k)u)^2} \tag{11}$$

where C_k^n is the Cotes coefficient, $C_k^{(n)} = \frac{1}{b-a} \int_{\tilde{a}}^{\tilde{b}} l_k(x)dx$, $l_k(x) = \int_{\tilde{a}}^{\tilde{b}} \prod_{j \neq k} \frac{(x-x_j)}{(x_k-x_j)} dx$.

Using the equipartition of nodes, the coordinates are transformed using $x = \tilde{a} + t\Delta h$. Using this transformation, the Cotes coefficients can be simplified further as:

$$C_k^{(n)} = \frac{\Delta h}{\tilde{b} - \tilde{a}} \int_0^n \prod_{\substack{j=0 \\ j \neq k}}^n \frac{(t-j)}{(k-j)} dt = \frac{(-1)^{n-k}}{k!(n-k)!} \frac{1}{n} \int_0^1 \prod_{\substack{j=0 \\ j \neq k}}^n (t-j) dt \tag{12}$$

Through Equations (10)–(12), the simplified mathematical equation can be obtained.

$$\ddot{u} + \mu\dot{u} + k_1 u - \alpha_2 k_3 u^3 = \frac{0.61\alpha_1}{(1-\delta\lambda-1.48u)^2} + 2\alpha_1 \rho \cos(\omega t) \frac{0.61}{(1-\delta\lambda-1.48u)^2}$$
$$+ \alpha_1 \rho^2 \cos^2(\omega t) \frac{0.61}{(1-\delta\lambda-1.48u)^2} \tag{13}$$

where $\delta = h/d$. It should be noted here that the maximum lateral displacement of the microbeam is at the midpoint viz., $y_{\max} = \phi(0.5)u \in [\lambda\delta, 1-\lambda\delta]$. At the middle point of microbeam, the value of the modal function is $\phi(0.5) = 1.59$. Therefore, the range of u is $u \in [\frac{\lambda\delta}{1.59}, \frac{1-\lambda\delta}{1.59}]$. The degree of matching is illustrated in Figure 2. The displacement is shown along the transverse coordinate and the electrostatic force along the ordinate.

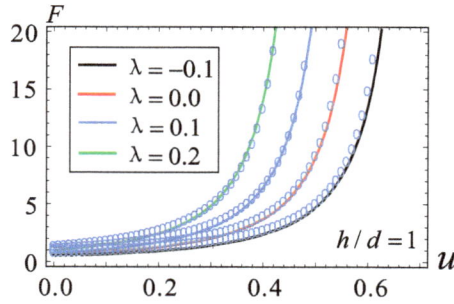

Figure 2. Contrast diagram of fitting curves under different section parameters. The circle is calculated using the numerical solution. The line is calculated using the Newton–Cotes method.

3. Static Analysis

The equilibrium position and the maximum pull-in voltage of MEMS can be found through static analysis. The geometric parameters of the microbeam are $L = 400$ μm, $b = 45$ μm, $d = 2$ μm, $h = 2$ μm, $E = 165$ GPa, $\rho = 2.33 \times 10^3$ kg/m³ and the dielectric constant in free space $\varepsilon_0 = 8.85 \times 10^{-12}$. Then, $-0.1 \leq \lambda \leq 0.69$ can be obtained from Equation (4). By removing the time-related items in Equation (13), the static response of the microresonator under the DC voltage actuation can be obtained.

$$k_1 u_s - \alpha_2 k_3 u_s^3 = \frac{0.61\alpha_1}{(1 - \delta\lambda - 1.48u_s)^2} \tag{14}$$

The relationship between the transverse displacement and DC voltage of the microbeam under different sections is shown in Figure 3a. With an increase of λ, the motion distance u_{s2} of the microbeam gradually decreases. However, u_{s2} has only a mathematical meaning and does not have any physical significance. In addition, the pull-in voltage and the pull-in location decreases as the λ increases. This is because the thickness of the beam increases as λ increases. As a result, the distance between the plate and the microbeam section decreases, which causes the axial movement distance to decrease resulting in pull-in. The influence of gap distance d on the static bifurcation is shown in Figure 3b. It can be seen from this figure that the equilibrium point u_{s2} is almost unchanged. The pull-in location increases slowly with the increase of d, and this effect is opposite that of λ. Therefore, the study of the relationship between λ and d is very necessary. The operating voltage range of the microresonator can be predicted through analysis. When $V_{DC} = 25$ V, several cases are selected to observe in Figure 4, and the yellow regions are the stable regions. It can be found that λ promotes the pull-in phenomenon, whereas d inhibits the pull-in occurrence. The results are consistent with the situation in Figure 3.

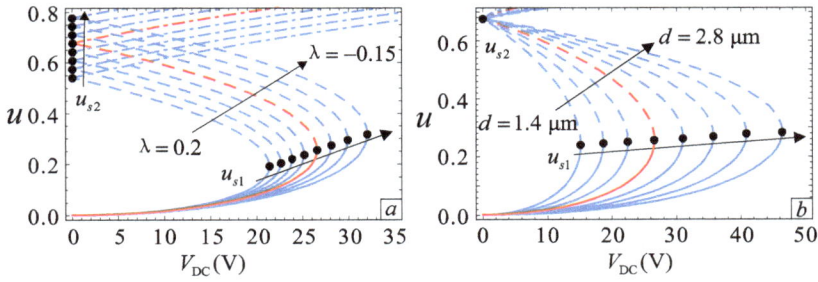

Figure 3. Relationship between the DC voltage and static equilibrium point under different physical parameters. (**a**) The influence of section parameters with $d = 2.0$ µm; (**b**) the influence of gap distance with $\lambda = 0$. The solid lines represent the stable solution. The dashed lines represent the unstable solution. The dotted lines are also stable, but it is impossible for them to appear in the physical model.

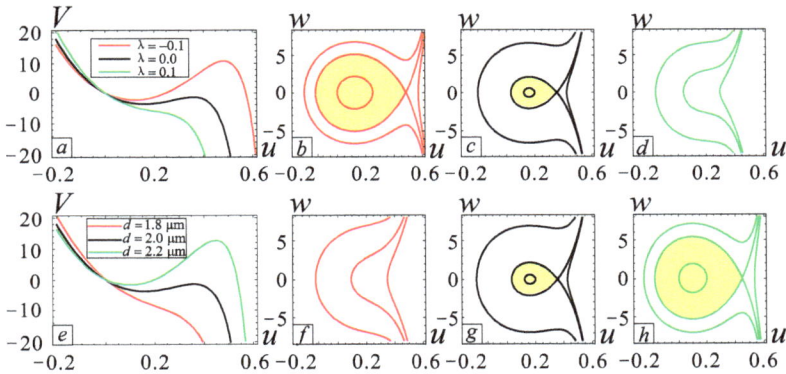

Figure 4. Potential energy curves and the corresponding phase diagrams under different physical parameters. (**a**) The potential energy curve under different section parameters with $d = 2.0$ µm; (**b–d**) are phase diagrams of $\lambda = -0.1$, $\lambda = 0$ and $\lambda = 0.1$, respectively; (**e**) the potential energy curve under different gap distances with $\lambda = 0$. (**f–h**) are the phase diagrams of $d = 1.8$ µm, $d = 2.0$ µm and $d = 2.2$ µm, respectively.

4. Dynamic Analysis

The resonance frequency and bifurcation behavior can be obtained through dynamic analysis. The MMS is used to investigate the response of the microresonator with small vibration amplitude around the stable equilibrium positions. Introducing $u = u_s + u_A$, u_s is the response to DC voltage and u_A is the response to AC voltage. The response of DC voltage u_s is obtained from Equation (14).

Substitute $u = u_s + u_A$ into Equation (13), and expand the electrostatic force equation up to third-order via Taylor expansion; the terms representing the equilibrium position can be eliminated. Since V_{AC} is far less than V_{DC} in the microresonator, the terms $V_{DC} = O(1)$ and $V_{AC} = O(\varepsilon^3)$ are considered. Here, ε is regarded as a small non-dimensional bookkeeping parameter. Therefore, Equation (13) can be modified as:

$$\ddot{u}_A + \varepsilon^2 \mu \dot{u}_A + \omega_n^2 u_A + a_q u_A^2 + a_c u_A^3 = \varepsilon^3 f \cos(\omega t) \qquad (15)$$

The symbolic meanings of ω_n a_q a_c and f are presented in Appendix A.

$$\omega = \omega_n + \varepsilon^2 \sigma \tag{16}$$

The approximate solution of Equation (15) can be obtained in the following form:

$$u_A(t, \varepsilon) = \varepsilon u_{A1}(T_0, T_1, T_2) + \varepsilon^2 u_{A2}(T_0, T_1, T_2) + \varepsilon^3 u_{A3}(T_0, T_1, T_2) \tag{17}$$

where $T_n = \varepsilon^n t$, $n = 0, 1, 2$.

Substituting Equations (16) and (17) into Equation (15) and equating the coefficients of like powers of ε, the following equations can be obtained:

$$O(\varepsilon^1) : D_0^2 u_{A1} + \omega_n^2 u_{A1} = 0 \tag{18}$$

$$O(\varepsilon^2) : D_0^2 u_{A2} + \omega_n^2 u_{A2} = -2D_0 D_1 u_{A1} - a_q u_{A1}^2 \tag{19}$$

$$O(\varepsilon^3) : \quad D_0^2 u_{A3} + \omega_n^2 u_{A3} = -2D_0 D_1 u_{A2} - 2D_0 D_2 u_{A1} - D_1^2 u_{A1} \\ -\mu D_0 u_{A1} - 2a_q u_{A1} u_{A2} - a_c u_{A1}^3 \\ +f \cos(\omega_n T_0 + \sigma T_2) \tag{20}$$

where $D_n = \frac{\partial}{\partial T_n}$, $n = 0, 1, 2$.

The general solution of Equation (18) can be written as:

$$u_{A1}(T_0, T_1, T_2) = A(T_1, T_2) e^{i\omega_n T_0} + \overline{A}(T_1, T_2) e^{-i\omega_n T_0} \tag{21}$$

Substituting Equation (21) into Equation (19), yields:

$$D_0^2 u_{A2} + \omega_n^2 u_{A2} = -2i\omega_n \frac{\partial A}{\partial T_1} e^{i\omega_n T_0} - a_q \left(A^2 e^{2i\omega_n T_0} + A\overline{A} \right) + cc \tag{22}$$

where cc represents the complex conjugate terms.

To eliminate the secular term, one needs:

$$-2i\omega_n \frac{\partial A}{\partial T_1} e^{i\omega_n T_0} = 0 \tag{23}$$

which indicates that A is only a function of T_2.

Thus, Equation (22) becomes:

$$D_0^2 u_{A2} + \omega_n^2 u_{A2} = -a_q \left(A^2 e^{2i\omega_n T_0} + A\overline{A} \right) + cc \tag{24}$$

The solution of u_{A2} can be given as:

$$u_{A2}(T_0, T_2) = \frac{a_q A^2}{3\omega_n^2} e^{2i\omega_n T_0} - \frac{a_q A\overline{A}}{\omega_n^2} + cc \tag{25}$$

Substituting Equations (21) and (25) into Equation (20) yields the secular terms:

$$2i\omega_n \frac{\partial A}{\partial T_1} + \mu i\omega_n A - \frac{10a_q^2 A^2 \overline{A}}{3\omega_n^2} + 3a_c A^2 \overline{A} - \frac{f}{2} e^{i\sigma T_2} = 0 \tag{26}$$

At this point, it is convenient to express A in the polar form:

$$A = \frac{1}{2} a(T_2) e^{i\beta(T_2)} + cc \tag{27}$$

Substituting Equation (27) into Equation (26) and separating the imaginary and real parts yield:

$$\frac{Da}{DT_2} = -\frac{\mu}{2}a + \frac{f}{2\omega_n}\sin\varphi \tag{28}$$

$$a\frac{D\varphi}{DT_2} = \sigma a + a^3\left(\frac{5a_q^2}{12\omega_n^3} - \frac{3a_c}{8\omega_n}\right) + \frac{f}{2\omega_n}\cos\varphi \tag{29}$$

where $\varphi = \sigma T_2 - \beta$.

The steady-state response can be obtained by imposing the conditions: $\frac{Da}{DT_2} = \frac{D\varphi}{DT_2} = 0$. Finally, the frequency response equation can be derived as follows:

$$a^2\left(\left(\frac{\mu}{2}\right)^2 + (\sigma + a^2\kappa)^2\right) = \left(\frac{f}{2\omega_n}\right)^2 \tag{30}$$

where $\kappa = \frac{5a_q^2}{12\omega_n^3} - \frac{3a_c}{8\omega_n}$.

The vibration peak value and backbone curve can be decided by $a_{max} = f/(\mu\omega_n)$ and $\omega = \omega_n - \kappa a_{max}$, respectively. The stability of the periodic solution can be determined by evaluating the eigenvalues of Jacobian matrix of Equations (28) and (29) at (a_0, φ_0).

$$J = \begin{vmatrix} -\frac{\mu}{2} & \frac{f}{2\omega_n}\cos\varphi_0 \\ 2\kappa a_0 - \frac{f}{2\omega_n a_0^2}\cos\varphi_0 & -\frac{f}{2\omega_n a_0}\sin\varphi_0 \end{vmatrix} \tag{31}$$

The system is stable if all the eigenvalues are negative; otherwise, the system is unstable [45].

The important dynamic properties of the microresonators include resonant frequency, frequency response, pull-in behavior, and so on. These properties show a very significant influence on the performance of microresonators. Therefore, the MMS and numerical analysis are used to observe the dynamic behaviors of microresonator under the influence of various factors such as different geometric parameters, external excitation, and so on.

4.1. Dynamic Analysis with Small Amplitude

The soft and hard nonlinearities of the system are related to k. Positive k can lead to a softening-type behavior, while the negative value can lead to a hardening-type behavior. Meanwhile, when the k value is approximately zero, i.e., the amplitude is small enough, the system experiences monostable vibration, i.e., linear-like vibration, which is an ideal state for MEMS designers. From Equation (30), one can notice that the nonlinear behaviors are affected by geometrical shape and electrostatic forces. The shape of the section is changed by adjusting the value of λ. Further, the relationship between the physical parameters and section shape is explored. The following physical quantities are assumed: $L = 400$ μm, $b = 45$ μm, $\rho = 2.33 \times 10^3$ kg/m^3, $E = 165$ GPa, dielectric constant $\varepsilon_0 = 8.85 \times 10^{-12}$ F/m and the clamped end thickness $h = 2$ μm. The other variation parameters are listed in Table 1. Next, the relationship between frequency response and various physical parameters is investigated.

Section changes due to the processing errors will have an impact on the system vibration. The effect of λ and d on the nonlinear behavior is shown in Figure 5, and the dimensionless damping coefficient $\mu = 0.1$. $V_{DC} = 20.5544$ V is obtained by calculating $\lambda = 0$ and $k = 0$. It can be noticed in Figure 5a,d that both λ and d can change the nonlinear soft and hard behavior. When $\lambda = -0.1$, the system has hard nonlinear characteristics. With the increase of λ, the system transits from hard to soft nonlinearity. $\lambda = 0$ is the dividing line. The parameters λ and d show an opposite effect. The points P_0, P_1, P_2, P_3 and P_4 are selected for analysis. Figure 5c,f shows the relationship between DC voltage and equivalent frequency at different λ and d. It can be seen that at $V_{DC} = 0$, the larger the value of λ is, the higher is the equivalent natural frequency. This is because of the increase of system stiffness, which is caused by the increase of λ. One can see from Figure 5c,f that the pull-in phenomenon could be promoted with an increase in the value of λ and a decrease in the value of d. At the same time, with the increase of the DC voltage, the equivalent frequency decreases. There is a significant "spring softening" phenomenon. The greater the value of λ is, the more obvious the phenomenon becomes.

Table 1. Simulation cases under different parameters.

Case	λ	d (μm)	Dynamic Behavior
P_1	-0.1	2	hardening-type vibration
P_0	0	2	linear-like vibration
P_2	0.1	2	softening-type vibration
P_3	0	1.8	softening-type vibration
P_4	0	2.0	hardening-type vibration

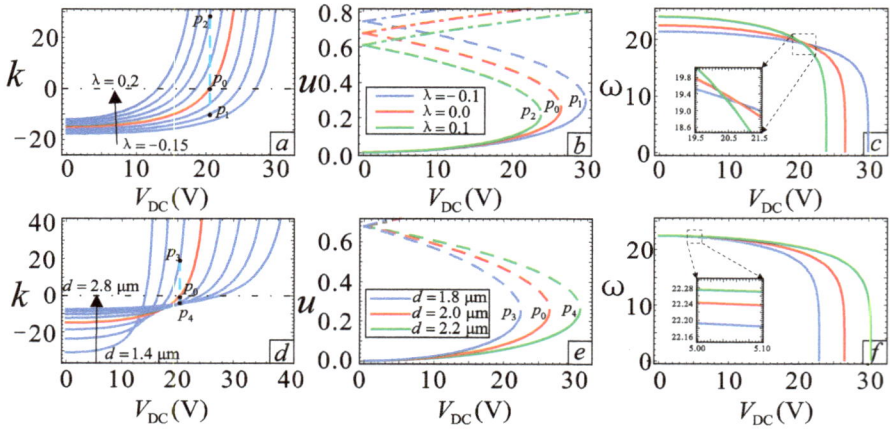

Figure 5. Relationship between DC voltage and mechanical behaviors under different physical parameters. (**a**–**c**) are the relationship between DC voltage and nondimensional parameter k, static equilibrium and equivalent natural frequency, respectively, under different section parameters with $d = 2.0$ μm. (**d**–**f**) are the relationship between DC voltage and nondimensional parameter k, static equilibrium and equivalent natural frequency, respectively, under different gap distance with $\lambda = 0$. The solid lines represent the stable solution. The dashed lines represent the unstable solution. The dotted lines also represent a stable case, but it is impossible for them to appear in the physical model.

To validate the above theoretical results, the frequency responses of five cases shown in Table 1 are studied using MMS. The long-time integration method of Equation (13) is used to obtain the numerical solutions. The accuracy of the results is verified by comparing both results. The AC excitation amplitude V_{AC} is varied to adjust the maximum amplitude. It can be known from Figure 6a–c that the nonlinear behavior changes from hardening to softening when $\lambda = -0.1$, $\lambda = 0$ and $\lambda = 0.1$. The system shows a hard nonlinearity behavior at $\lambda = -0.1$. When $\lambda = 0.1$, the system shows a soft nonlinearity behavior, and $\lambda = 0$ is the dividing line, where the vibration is linear.

In addition to λ, the gap distance d will also affect the frequency response. The parameters λ and d affect the soft and hard behavior in the opposite way. The nonlinear behavior changes from soft nonlinear to hard nonlinear when $d = 1.8$ μm, $d = 2.0$ μm and $d = 2.2$ μm, as shown in Figure 6a,d,e. Therefore, adjusting the relationship between d and λ to achieve linear behavior is necessary.

However, this time, the numerical and analytical solutions do not match at point P_0. The numerical solution shows a softening behavior, whereas the analytical solution shows a linear behavior. This situation will be elaborated in detail below.

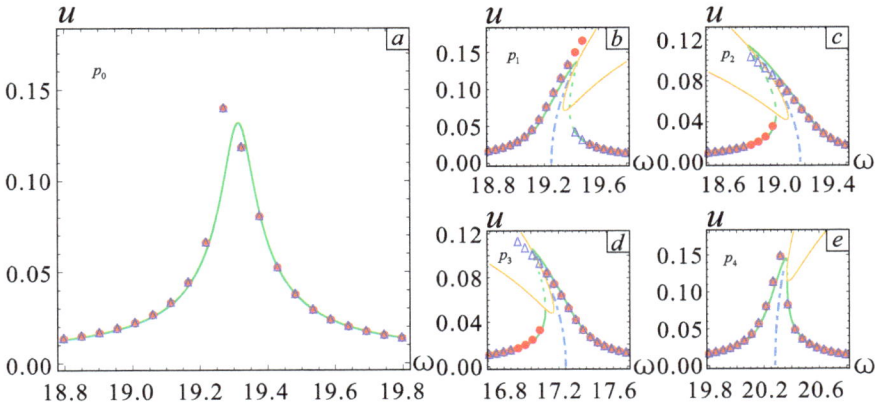

Figure 6. Frequency response curve in different situations, i.e., corresponding to cases P_0, P_1, P_2, P_3 and P_4. (**a**) is the case of P_0. (**b**) is the case of P_1. (**c**) is the case of P_2. (**d**) is the case of P_3. (**e**) is the case of P_4. The solid lines represent stable solutions. The dashed lines represent unstable solutions. The point represents the numerical solution.

4.2. Dynamic Analysis with Large Amplitude

With an increase of AC voltage, the MEMS resonator may undergo large amplitude vibration. When the AC voltage is increased to beyond a certain value, the analytical and numerical solutions will not match, for example, corresponding to the point P_0 in Figure 6a. This phenomenon will be analyzed in detail in Figure 7. In addition, the $V_{DC} = 15$ V and $V_{DC} = 23$ V cases are considered to observe the phenomenon in the soft and hard nonlinearity cases. The amplitude of vibration increases, i.e., shifts from left to right, when the AC voltage is adjusted. When the vibration amplitude is small, the numerical and analytical solutions match very well. When the amplitude is close to $u = 0.2$, this difference between the solutions begins to appear. This phenomenon shows that the softening effect of analytic solutions is weakened. This is because the higher order terms in the Taylor expansion of the electrostatic force equation are omitted during the simplification process. These higher order terms are negligible when the amplitude is small. However, as the amplitude increases, these terms are not negligible; especially the frequency response in the red frame, which transits from hardening to softening behavior. When the vibration amplitude reaches around $u = 0.3$ and as the DC voltage increases, the influence of electrostatic force nonlinearity exceeds the structural stiffness nonlinearity. At this time, the electrostatic force plays a leading role. During sweep frequency response analysis, the resonator at the point of P may generate two kinds of motion. The motion is dynamic pull-in instability or jumping motion to the upper stable branch.

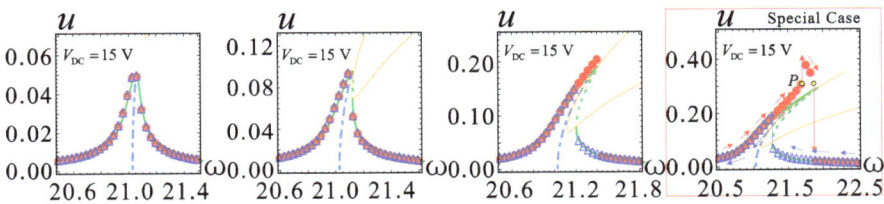

Figure 7. *Cont.*

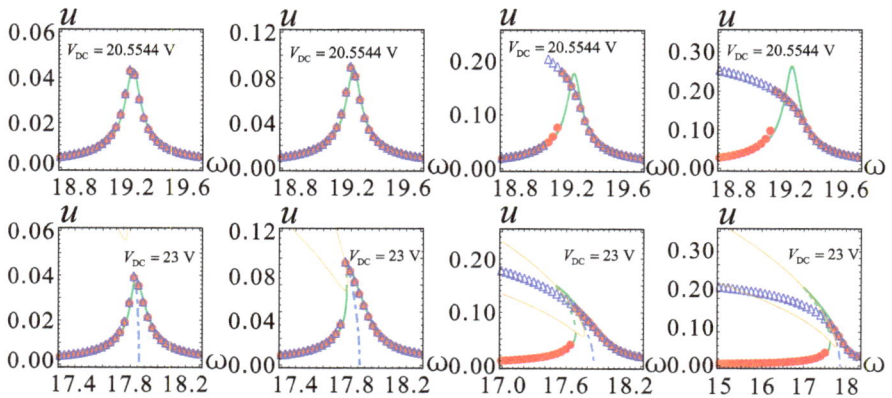

Figure 7. The frequency response changes with AC voltage at $\lambda = 0$. The DC voltages are $V_{DC} = 15$ V, $V_{DC} = 20.5544$ V and $V_{DC} = 23$ V, respectively, from top to bottom in the figure. The solid lines represent stable solutions. The dashed lines represent unstable solutions. The point represents a numerical solution.

The frequency response in the red frame is further considered for a detailed analysis; similarly, the two models for the $\lambda = -0.1$ and $\lambda = 0.1$ cases are selected for analysis. It can be seen from Figure 8 that in the case of section parameter $\lambda = 0$, when the AC voltage amplitude $V_{AC} \geq 0.22$ V, the system shows a jump in the frequency response. The vibration amplitude becomes large with the increase of V_{AC}, but the jump point does not change. The effect of the section parameter on jump phenomena is shown in Figure 9. The increase of λ will promote the occurrence of the jump phenomenon. On the contrary, when λ is small, a higher voltage is needed to observe a similar behavior. At the same time, the system will generate more energy output.

The frequency is selected near the jump point in each case, and the corresponding time history curves are shown in Figures 9–11. By setting different initial value x_0, the displacement of all stable solutions can be obtained. If there is no hardening-to-softening behavior, the vibration of the resonator will appear from one stable solution to two stable solutions in the vicinity of the jump point *SN1* as shown in Figure 9. The case of two stable solutions appears after jump point *SN1*. If the hardening-to-softening behavior appears, the solution case is the same as the one shown in Figure 9 when the nonlinearity is weak. On the contrary, in this situation, the two stable solutions case appears before the jump point *SN1*, as shown in Figure 10. When the nonlinearity is strong as depicted in Figure 11, there will be three stable solutions at most, and it changes into two stable solutions after the jump at the *SN2* point. As the frequency increases, the stable solution finally returns to one.

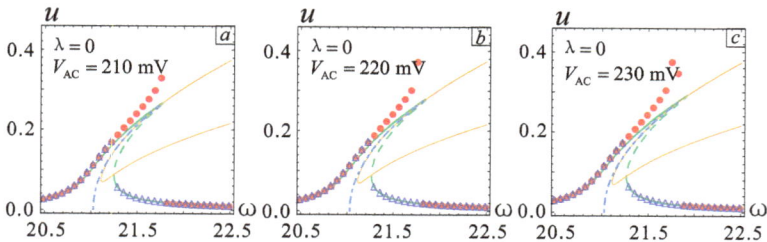

Figure 8. The changes in frequency response as AC voltage increases at $\lambda = 0$. (**a**) is the case of $V_{AC} = 210$ mV. (**b**) is the case of $V_{AC} = 220$ mV. (**c**) is the case of $V_{AC} = 230$ mV. The solid lines represent stable solutions. The dashed lines represent unstable solutions. The point represents the numerical solution.

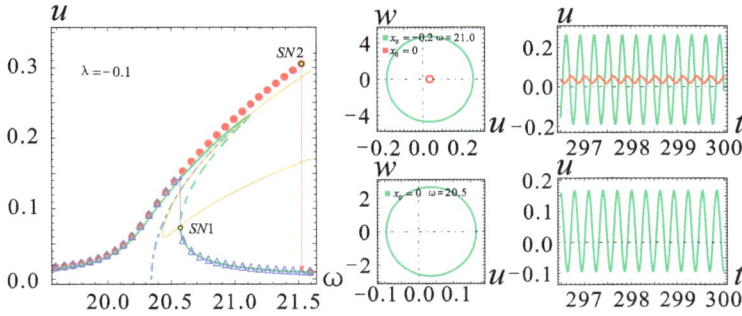

Figure 9. Phase diagram and the corresponding time history curve at $\lambda = -0.1$. The solid lines represent stable solutions. The dashed lines represent unstable solutions. The point represents the numerical solution.

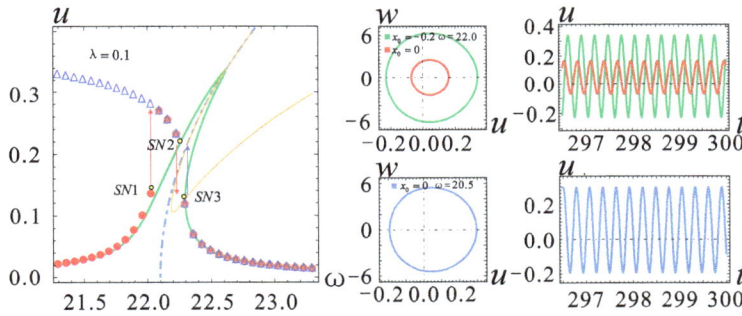

Figure 10. Phase diagram and the corresponding time history curve at $\lambda = 0.1$. The solid lines represent stable solutions. The dashed lines represent unstable solutions. The point represents the numerical solution.

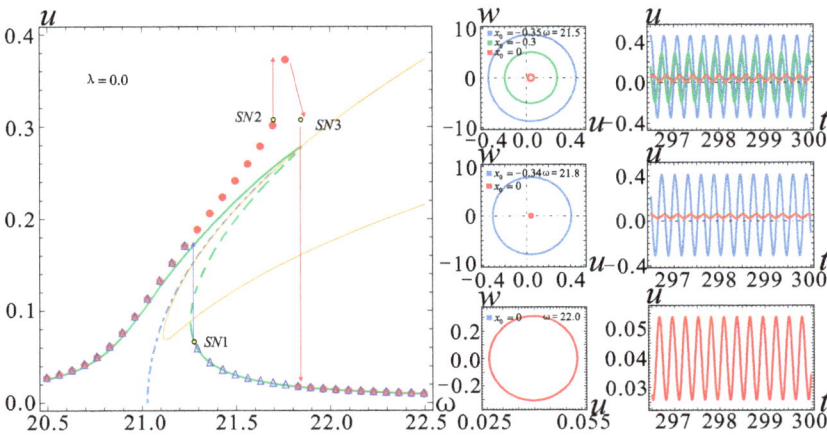

Figure 11. Phase diagram and the corresponding time history curve at $\lambda = 0$. The solid lines represent stable solutions. The dashed lines represent unstable solutions. The point represents the numerical solution.

5. Finite Element Verification

Static and dynamic analyses of the system were carried out through the mathematical model in the previous research. The influence of surface machining error and gap distance on the nonlinear vibration was obtained. However, the results are obviously not convincing since the analysis was carried out using the mathematical model alone. In this section, the following physical quantities are assumed: $L = 400$ μm, $b = 45$ μm, $d = 2$ μm, $\rho = 2.33 \times 10^3$ kg/m^3, $E = 165$ GPa, dielectric constant $\varepsilon_0 = 8.85 \times 10^{-12}$ F/m and the clamped end thickness $h = 2$ μm. The finite element simulations of the several λ values are carried out using COMSOL software. The module used for this analysis is the MEMS module, and the electrical physical field interface is selected. The interface combines solid mechanics and electrostatics with the dynamic grids to model the deformation of an electrostatically-actuated structure. The number of degrees of freedom for solving this system is 31,185. Some nonlinear behaviors such as the pull-in effect and the electrostatic force softening effect are simulated in the real situation. The simulation of the pull-in voltage is carried out in the steady-state solver. The simulation of electrostatic force softening first proceeds through the parameterized scanning of DC voltage and then calculates the corresponding value of each point voltage in the steady-state solver and the eigenvalue solver. The physical model established through the finite element software is shown in Figure 12. As shown in the figure, below is the beam model and above is the air area.

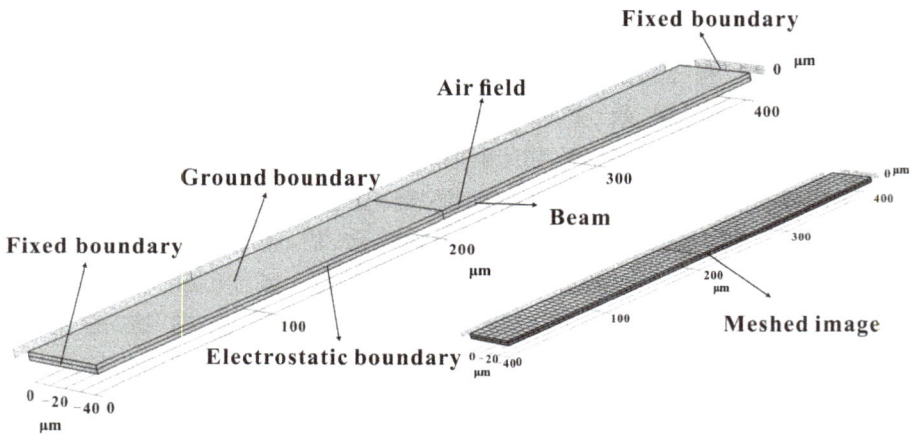

Figure 12. COMSOL simulation model.

A comparison between the finite element simulation and analytical solution on the electrostatic softening effect is shown in Figure 13. Before $V_{DC} = 20$ V, the two results are in the good agreement. However, the error starts to increase near the pull-in position when the DC voltage exceeds 20 V. It is evident from these results that the error increases as the value of λ increases. The finite element simulation and analytical solution of the static pull-in effect are shown in Figure 14. It can be found that the pull-in position of the two results is same. The pull-in voltage is well simulated at $\lambda = 0$, while the other two cases are slightly different. It can be seen from Figures 13 and 14 that the maximum error occurs near the pull-in point. The reason for the error could be as follows: the COMSOL software acquiescent structure stiffness is linear, while the actual system contains nonlinear stiffness. Although the analytical solutions take the nonlinear factors into account in the analysis, because of the limitation of MMS, an error between the analytical and numerical solution is inevitable especially when the amplitude is too large. The comprehensive mechanical behavior of the system cannot be obtained only through the numerical method. Therefore, the contradiction really needs further consideration, which is not within the scope of this paper.

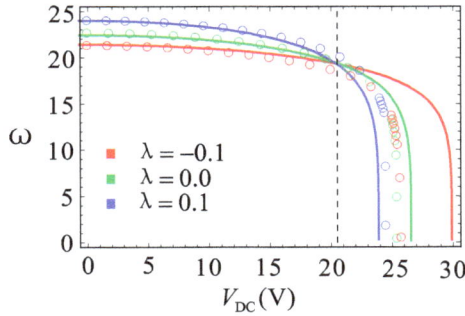

Figure 13. Relationship between DC voltage and equivalent frequency under different section parameters. The solid lines represent the analytic solutions. The circles represent the finite element solutions.

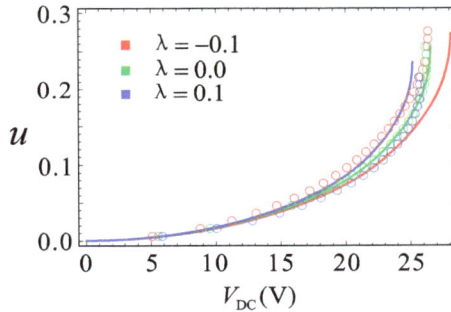

Figure 14. Relationship between DC voltage and static equilibrium point under different section parameters. The solid lines represent the analytic solutions. The circles represent the finite element solutions.

6. Conclusions

In this paper, the static and dynamic characteristics of a doubly-clamped electrostatic microresonator considering the effect of surface processing error on the thickness are studied. A section parameter is proposed to describe the microbeam changes in the upper and lower sections. The Galerkin discrete method is used to decouple, and the finite element analysis is carried out using the software COMSOL. To observe the system's nonlinear vibration, the MMS is applied to obtain the approximate frequency response equation, and the long-time integral method is used to verify. The key conclusions are as follows.

(1) From derivation, the range of section parameter in micro resonators is $\lambda \in [-0.3, 0.9]$.

(2) The occurrence of pull-in phenomenon could be promoted by λ increasing and d decreasing. Several typical cases are analyzed by using the potential energy curve and phase diagram. With either the increase of the parameter λ or the decrease of d, the barrier energy gradually decreases and the safe region reduces. As a result, the pull-in will occur.

(3) Under small perturbations, the resonator may vibrate in the neighborhood of the equilibrium point. When the gap distance is constant, the sectional parameter $\lambda > 0$ will make the system vibration tend to softening-type behavior. On the contrary, $\lambda < 0$ will make the system vibration lean towards hardening-type behavior. When the section parameter is constant, as the gap distance of the microbeam is larger, the hardening-type behavior more easily appears. Similarly, as the gap distance is smaller, the softening-type behavior is easier to obtain. Therefore, if the

microresonator is thinner or thicker because of the surface machining error, the gap distance *d* can be adjusted to make the system vibration close to linear.

(4) The frequency response is obtained by MMS will lead to the nonlinear softening effect being weakened. This error is negligible when the amplitude of vibration is relatively small. As the amplitude increases beyond a certain value, this error will be more obvious. If the nonlinearity exhibits hardening-type behavior at the beginning, the nonlinearity of electrostatic force will gradually strengthen with the increases of the amplitude. Finally, the electrostatic force began to dominate when its nonlinearity effect on the system exceeded the influence of structural stiffness nonlinearity. At this time, the frequency response will exhibit hardening to softening behavior. The higher the value of λ is, the more easily it appears.

It can be concluded from the presented results that the surface processing error does affect the static and dynamic characteristics of the microresonator. When the existing micromachining process is not improved, it will go for a revision only after considering the processing errors in the original theoretically-based design. It can make the final product meet the theoretical design requirements and increase the rate of finished products.

Acknowledgments: This project was supported by the National Natural Science Foundation of China (Grant No. 11602169, No. 11772218, No. 11702192 and No. 51605330) and the Natural Science Foundation of Tianjin City (No. 16JCQNJC04700 and No. 17JCYBJC18900).

Author Contributions: Jingjing Feng, Cheng Liu and Wei Zhang conceived and designed the model; Jingjing Feng and Cheng Liu contributed theoretical analysis; Jingjing Feng and Shuying Hao analyzed the data; Cheng Liu, Wei Zhang and Shuying Hao conducted the simulation; Cheng Liu and Jingjing Feng wrote the paper.

Conflicts of Interest: The authors declare no conflict of interest.

Appendix A

$$g = \int_0^1 A(x)\phi^2(x)dx \tag{A1}$$

$$\mu = c/g \tag{A2}$$

$$k_1 = \int_0^1 (I(x)\phi''(x))''\phi(x)dx/g \tag{A3}$$

$$k_3 = \int_0^1 A(x)(\phi'(x))^2 dx \int_0^1 \phi''(x)\phi(x)dx/g \tag{A4}$$

$$\omega_n = \sqrt{k_1 - 3\alpha_2 k_3 u_s^2 - \frac{1.8056\alpha_1}{(1 - 1.48u_s - \delta\lambda)^3}} \tag{A5}$$

$$a_q = -3\alpha_2 k_3 u_s - \frac{4.00843\alpha_1}{(1 - 1.48u_s - \delta\lambda)^4} \tag{A6}$$

$$a_c = -\alpha_2 k_3 - \frac{7.90997\alpha_1}{(1 - 1.48u_s - \delta\lambda)^5} \tag{A7}$$

$$f = \frac{1.22\alpha_1\rho}{(1 - 1.48u_s - \delta\lambda)^2} \tag{A8}$$

References

1. Younis, M.I.; Nayfeh, A.H. A study of nonlinear response of a resonant microbeam to an electric actuation. *Nonlinear Dyn.* **2003**, *31*, 91–117. [CrossRef]
2. Leus, V.; Elata, D. On the dynamic response of electrostatic MEMS switches. *J. Microelectromech. Syst.* **2008**, *17*, 236–243. [CrossRef]

3. Lin, W.H.; Zhao, Y.P. Casimir effect on the pull-in parameters of nanometer switches. *Microsyst. Technol.* **2005**, *11*, 80–85. [CrossRef]
4. Nayfeh, A.H.; Ouakad, H.M.; Najar, F.; Choura, S.; Abdel-Rahman, E.M. Nonlinear dynamics of a resonant gas sensor. *Nonlinear Dyn.* **2009**, *59*, 607–618. [CrossRef]
5. Ibrahim, A.; Younis, M.I. Simple fall criteria for MEMS sensors: Data analysis and sensor concept. *Sensors* **2014**, *14*, 12149–12173. [CrossRef] [PubMed]
6. Zaitsev, S.; Shtempluck, O.; Buks, E.; Gottlieb, O. Nonlinear damping in a micromechanical oscillator. *Nonlinear Dyn.* **2011**, *67*, 859–883. [CrossRef]
7. Nayfeh, A.H.; Younis, M.I.; Abdel-Rahman, E.M. Dynamic pull-in phenomenon in MEMS resonators. *Nonlinear Dyn.* **2006**, *48*, 153–163. [CrossRef]
8. Zhang, W.M.; Tabata, O.; Tsuchiya, T.; Tsuchiya, T.; Meng, G. Noise-induced chaos in the electrostatically actuated MEMS resonator. *Phys. Lett. A* **2011**, *375*, 2903–2910. [CrossRef]
9. Al Hafiz, M.A.; Kosuru, L.; Ramini, A.; Chappanda, K.N.; Younis, M.I. In-plane MEMS shallow arch beam for mechanical memory. *Micromachines* **2016**, *7*, 191. [CrossRef]
10. Zhang, W.M.; Yan, H.; Peng, Z.K.; Meng, G. Electrostatic pull-in instability in MEMS/NEMS: A review. *Sens. Actuators A Phys.* **2014**, *214*, 187–218. [CrossRef]
11. Ibrahim, A.; Jaber, N.; Chandran, A.; Thirupathi, M.; Younis, M. Dynamics of microbeams under multi-frequency excitations. *Micromachines* **2017**, *8*, 32. [CrossRef]
12. Haluzan, D.T.; Klymyshyn, D.M.; Achenbach, S.; Borner, M. Reducing pull-in voltage by adjusting gap shape in electrostatically actuated cantilever and fixed-fixed beams. *Micromachines* **2010**, *1*, 68–81. [CrossRef]
13. Fang, Y.M.; Li, P. A new approach and model for accurate determination of the dynamic Pull-in parameters of microbeams actuated by a step voltage. *J. Micromech. Microeng.* **2013**, *23*, 109501. [CrossRef]
14. Abdel-Rahman, E.M.; Younis, M.I.; Nayfeh, A.H. Characterization of the mechanical behavior of an electrically actuatedmicrobeam. *J. Micromech. Microeng.* **2002**, *12*, 759–766. [CrossRef]
15. Younis, M.I.; Abdel-Rahman, E.M.; Nayfeh, A. Static and dynamic behavior of an electrically excited resonant microbeam. In Proceedings of the 43rd AIAA Structures, Structural Dynamics, and Materials Conference, Denver, CO, USA, 22–25 April 2002; pp. 1298–1305.
16. Alsaleem, F.M.; Younis, M.I. Stabilization of electrostatic MEMS resonators using a delayed feedback controller. *Smart Mater. Struct.* **2010**, *19*, 035016. [CrossRef]
17. Wang, L.; Hong, Y.Z.; Dai, H.L.; Ni, Q. Natural frequency and stability tuning of cantilevered cnts conveying fluid in magnetic field. *Acta Mech. Solida Sin.* **2016**, *59*, 567–576. [CrossRef]
18. Lv, H.W.; Li, Y.H.; Li, L.; Liu, Q.K. Transverse vibration of viscoelastic sandwich beam with time-dependent axial tension and axially varying moving velocity. *Appl. Math. Model.* **2014**, *38*, 2558–2585. [CrossRef]
19. Ghayesh, M.H.; Farokhi, H.; Amabili, M. Nonlinear behaviour of electrically actuated MEMS resonators. *Int. J. Eng. Sci.* **2013**, *71*, 137–155. [CrossRef]
20. Zhang, M.W.; Meng, G. Nonlinear dynamic analysis of electrostatically actuated resonant MEMS sensors under parametric excitation. *IEEE. Sens. J.* **2007**, *7*, 370–380. [CrossRef]
21. Ibrahim, M.I.; Younis, M.I.; Alsaleem, F. An investigation into the effects of electrostatic and squeeze-film non-linearities on the shock spectrum of microstructures. *Int. J. Non-Linear Mech.* **2010**, *45*, 756–765. [CrossRef]
22. Ghayesh, M.H.; Farokhi, H. Nonlinear dynamics of microplates. *Int. J. Eng. Sci.* **2015**, *86*, 60–73. [CrossRef]
23. Gholipour, A.; Farokhi, H.; Ghayesh, M.H. In-plane and out-of-plane nonlinear size-dependent dynamics of microplates. *Nonlinear Dyn.* **2015**, *79*, 1771–1785. [CrossRef]
24. Alsaleem, F.M.; Younis, M.I.; Ruzziconi, L. An experimental and theoretical investigation of dynamic pull-in in MEMS resonators actuated electrostatically. *J. Microelectromech. Syst.* **2010**, *19*, 794–806. [CrossRef]
25. Alkharabsheh, S.A.; Younis, M.I. Statics and dynamics of MEMS arches under axial forces. *J. Vib. Acoust.* **2013**, *135*, 021007. [CrossRef]
26. Krylov, S.; Harari, I.; Cohen, Y. Stabilization of electrostatically actuated microstructures using parametric excitation. *J. Micromech. Microeng.* **2005**, *15*, 1188–1204. [CrossRef]
27. Najar, F.; Choura, S.; Abdel-Rahman, E.M. Dynamic analysis of variable-geometry electrostatic microactuators. *J. Micromech. Microeng.* **2006**, *16*, 2449–2457. [CrossRef]
28. Younis, M.I.; Ouakad, H.M.; Alsaleem, F.M.; Miles, R.; Cui, W.L. Nonlinear dynamics of MEMS arches under harmonic electrostatic actuation. *J. Microelectromech. Syst.* **2010**, *19*, 647–656. [CrossRef]

29. Farokhi, F.; Ghayesh, M.H.; Amabili, M. Nonlinear dynamics of a geometrically imperfect microbeam based on the modified couple stress theory. *Int. J. Eng. Sci.* **2013**, *68*, 11–23. [CrossRef]
30. Farokhi, H.; Ghayesh, M.H. Nonlinear dynamical behaviour of geometrically imperfect microplates based on modified couple stress theory. *Int. J. Mech. Sci.* **2015**, *90*, 133–144. [CrossRef]
31. Farokhi, H.; Ghayesh, M.H. Thermo-mechanical dynamics of perfect and imperfect Timoshenko microbeams. *Int. J. Eng. Sci.* **2015**, *91*, 12–33. [CrossRef]
32. Ruzziconi, L.; Younis, M.I.; Lenci, S. Parameter identification of an electrically actuated imperfect microbeam. *Int. J. Non-Linear Mech.* **2013**, *57*, 208–219. [CrossRef]
33. Ruzziconi, L.; Younis, M.I.; Lenci, S. An electrically actuated imperfect microbeam: Dynamical integrity for interpreting and predicting the device response. *Meccanica* **2013**, *48*, 1761–1775. [CrossRef]
34. Krylov, S.; Ilic, B.R.; Schreiber, D. The pull-in behavior of electrostatically actuated bistable microstructures. *J. Micromech. Microeng.* **2008**, *18*, 055026. [CrossRef]
35. Xu, T.T.; Ruzziconi, L.; Younis, M.I. Global investigation of the nonlinear dynamics of carbon nanotubes. *Acta Mech.* **2017**, *228*, 1029–1043. [CrossRef]
36. Ouakad, H.M.; Younis, M.I. The dynamic behavior of MEMS arch resonators actuated electrically. *Int. J. Nonlinear Mech.* **2010**, *45*, 704–713. [CrossRef]
37. Mobki, H.; Rezazadeh, G.; Sadeghi, M.; Vakili-Tahami, F.; Seyyed-Fakhrabadi, M.M. A comprehensive study of stability in an electro-statically actuated micro-beam. *Int. J. Non-Linear Mech.* **2013**, *48*, 78–85. [CrossRef]
38. Han, J.X.; Qi, H.J.; Gang, J.; Li, B.Z.; Feng, J.J.; Zhang, Q.C. Mechanical behaviors of electrostatic microresonators with initial offset imperfection: Qualitative analysis via time-varying capacitors. *Nonlinear Dyn.* **2017**, 1–27. [CrossRef]
39. Herrera-May, A.L.; Aguilera-Cortés, L.A.; García-Ramírez, P.J.; Plascencia-Mora, H.; Torres-Cisneros, M. Modeling of the intrinsic stress effect on the resonant frequency of NEMS resonators integrated by beams with variable cross-section. *Microsyst. Technol.* **2010**, *16*, 2067–2074. [CrossRef]
40. Joglekar, M.M.; Pawaskar, D.N. Shape optimization of electrostatically actuated microbeams for extending static and dynamic operating ranges. *Struct. Multidiscip. Optim.* **2012**, *46*, 871–890. [CrossRef]
41. Trivedi, R.R.; Joglekar, M.M.; Shimpi, R.R.; Pawaskar, D.N. Shape optimization of electrostatically actuated micro cantilever beam with extended travel range using simulated annealing. *Lect. Notes Eng. Comput. Sci.* **2011**, *2192*, 2042–2047.
42. Zhang, S.; Zhang, W.M.; Peng, Z.K.; Meng, G. Dynamic characteristics of electrostatically actuated shape optimized variable geometry microbeam. *Shock Vib.* **2015**, 867171. [CrossRef]
43. Kuang, J.H.; Chen, C.J. Dynamic characteristics of shaped microactuators solved using the differential quadrature method. *J. Micromech. Microeng.* **2004**, *14*, 647–655. [CrossRef]
44. Najar, F.; Choura, S.; El-Borgi, S.; Abdel-Rahman, E.M.; Nayfeh, A.H. Modeling and design of variable-geometry electrostatic microactuators. *J. Micromech. Microeng.* **2005**, *15*, 419–429. [CrossRef]
45. Shao, S.; Masri, K.M.; Younis, M.I. The effect of time-delayed feedback controller on an electrically actuated resonator. *Nonlinear Dyn.* **2013**, *74*, 257–270. [CrossRef]

micromachines

MDPI

Article

The Application of Chemical Foaming Method in the Fabrication of Micro Glass Hemisphere Resonator

Jianbing Xie *, Lei Chen, Hui Xie, Jinqiu Zhou and Guangcheng Liu

Key Laboratory of Micro/Nano Systems for Aerospace, Ministry of Education,
Northwestern Polytechnical University, Xi'an 710072, China; cl@mail.nwpu.edu.cn (L.C.);
hhx@mail.nwpu.edu.cn (H.X.); zjq523543302@mail.nwpu.edu.cn (J.Z.); lgc1995@mail.nwpu.edu.cn (G.L.)
* Correspondence: xiejb@nwpu.edu.cn; Tel.: +86-29-8846-0353

Received: 20 November 2017; Accepted: 19 January 2018; Published: 24 January 2018

Abstract: Many researchers have studied the miniaturization of the hemisphere resonator gyroscope for decades. The hemisphere resonator (HSR), as the core component, has a size that has been reduced to the submillimeter level. We developed a method of batch production of micro-hemisphere shell resonators based on a glass-blowing process to obtain larger hemisphere shells with a higher ratio of height to diameter (H/D), we introduced the chemical foaming process (CFP) and acquired an optimized hemisphere shell; the contrasted and improved H/D of the hemisphere shell are 0.61 and 0.80, respectively. Finally, we increased the volume of glass shell resonator by 51.48 times while decreasing the four-node wineglass resonant frequencies from 7.24 MHz to 0.98 MHz. The larger HSR with greater surface area is helpful for setting larger surrounding drive and sense capacitive electrodes, thereby enhancing the sensitivity of HSR to the rotation. This CFP method not only provides more convenience to control the shape of a hemisphere shell but also reduces non-negligible cost in the fabrication process. In addition, this method may inspire some other research fields, e.g., microfluidics, chemical analysis, and wafer level package (WLP).

Keywords: hemisphere resonator gyroscope; hemisphere resonator; chemical foaming process; glassblowing; hemisphere shell; hollow glass microsphere; micro electro mechanical systems (MEMS)

1. Introduction

The hemisphere resonator gyroscope (HRG), as one type of Coriolis gyro, has superior navigation precision because of its unique axisymmetric 3D structure [1]. Using a MEMS-based process, many different mHRGs and uHRGs have already been studied by global research teams. The different structural designs of the resonator mainly include hollow hemisphere [2], wineglass [3], disk [4], these resonators are usually surrounded by several coupled drive and sense electrodes. The fabrication processes of the resonator also differ, including deep reactive ion etching (DRIE), glassblowing [5], atomic layer deposition [6], sacrificial layer etching [7], or a combination of the above methods.

Quality factor (Q-factor) is a dimensionless characteristic, which is defined as the ratio of the energy stored to the power loss in every cycle when a system works on resonant status. The hemisphere resonator (HSR) with a high Q is desired because it can reduce mechanical noise. The Q factor of the HSR mainly depends on thermoelastic damping and anchor losses, studies of the detailed energy loss mechanisms were presented in [7,8]. A low thermoelastic damping structural material can achieve a high Q factor; for example, an HSR fabricated by microcrystalline diamond was found to have a four-node wineglass resonant frequency of 18.321 kHz, with the observed Q of ~10,000 at the four-node wineglass mode and 20,000 at six-node vibration mode [9]. A stem-supported hemispherical shell with self-aligned tall capacitive electrodes was fabricated in polysilicon; the shell has a wineglass resonant mode at 5.58 kHz with a Q of 17,600 [10]. Other materials, such as fused silica [11] and Pyrex have also been used to fabricate an HSR [2]. The ideal drive axis and sense axis of a gyroscope should be mutually perpendicular, which is called quadrature coupling, but the fabrication imperfections, such as lithographic misalignment and inhomogeneous anchor region, would prevent the angle of drive axis and sense axis from being 90° and result in quadrature error. When compared with the conventional 2D flat massive block typed gyroscope, the inherent central symmetry resonator can reduce quadrature error and provide more flexible signal excitation and pick up design.

A micro-hemisphere glass shell resonator is a good choice as the central component of the HRG. Figure 1a shows a typical structure demonstration of a glass HSR, which has many advantages, such as high stiffness, good shock resistance, and low thermal expansion coefficient consistent with that of a silicon substrate [12]. The silicon substrate also provides high aspect ratio surrounding electrodes and enables simultaneous fabrication [13]. Moreover, the processes for batch fabrication this type HSR have been developed in [5], the key processes (which include DRIE, anodic bonding, and glass blowing) are simple and convenient techniques those are easily accessible to related research groups. For example, a 500-μm radius spherical resonator with a four-node wineglass resonant frequency of 1.37 MHz and a Q factor of 1280 in vacuum (0.4 mT) was reported in [6]. Further Q enhancement is possible by minimizing the shell anchoring to the substrate or by using high-Q materials [2].

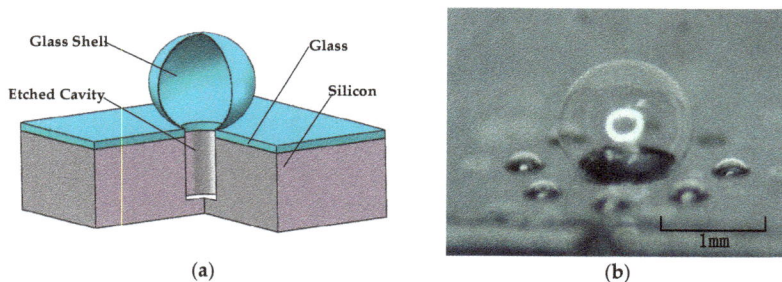

Figure 1. Demonstration diagram and micrograph of the glass hemisphere resonator (HSR): (**a**) cross-sectional view of the micro glass shell structure; (**b**) microphotograph of a glass-blown shell (height/diameter (H/D) = 0.79).

At the beginning, we utilized a similar process to fabricate the micro hemispherical glass shell, however, the shell shapes were not very uniform. Figure 1b shows the highest hemisphere glass shell blown by an 800 μm deep cavity, the radius of the etched cavity is 250 μm. We measured the shape of the glass HSR via scanning electron microscope (SEM), the radius of this HSR is 439.18 μm. We calculated that the glass HSR has a ratio of height to diameter (H/D) of 0.79. However, it is difficult to obtain such a large glass shell because the H/D of most other samples did not exceed 0.5. An HSR with high H/D is desirable for the larger HSR has several advantages, for example, greater surface area is helpful for setting higher aspect surrounding drive and sense capacitive electrodes, which can increase the sensitivity of HRG, and HSR with high H/D would lower its four-node wineglass resonant frequency, which is beneficial for the drive and sense of the HRG. Finally, we managed to develop an available method to produce larger, uniform glass HSRs.

There are some challenges in the fabrication of a glass HSR, for example, very deep cylindrical cavities (usually 800 μm in depth) must be etched previously, requiring a thick silicon substrate (equal or greater than 1000 μm in thickness). This thickness requirement not only increases the process time and cost but also influences the subsequent processes because the excessively thick silicon substrate increases the difficulty of the anodic bonding and dicing processes. Even with the use of such a deep cavity to blow a hemispheric glass shell, the volume of the glass shell is still under restrictions from the process parameters; thus, we seek a larger hemisphere glass shell by improving the process condition or introducing some other valid approaches. To get a micro glass hemisphere shell with higher H/D, Andrei M. Shkel et al. presented an alternative fabrication process to make an extra trapped air pocket [2,5], utilizing two bonded silicon wafers to achieve larger sphericity (H/D), because which can provide more volume of the sealed gas. D. Senkal et al. also utilized an additional stencil silicon layer to create an air pocket and fabricated a spherical shell structure with a stem, which was conventional micro machining technology [14]. Jintang Shang et al. used foaming agent TiH_2 to fabricate glass bubbles [15] and hemispherical glass shells with self-formed stems [16]. To obtain a larger glass HSR, we managed to introduce chemical foaming process (CFP) method in the fabrication of glass-blown shell resonators, the H/D of glass hemisphere shell blown by 200 µm deep etched cavity with quantified foaming agent can approach (even exceed) the HSR blown by the 800 µm deep etched cavity without foaming agent.

In this work, analytical model and experimental verification were presented sequentially. We utilized the height model to predict the shape of the glass-blown spherical shell, and then correlative bonded stacks were heated in the furnace at a temperature above the softening point of the glass. In the initial experiment stage, the resulting shapes were almost hemispheric shells (Figure 2a), and the H/D of the shell samples was 0.61. We want to acquire a hollow glass shell that is closer to a sphere, because this structure could provide more desirable properties of the gyroscope. Finally, we explored an available method of transferring quantified foaming agent into the etched cavity before anodic bonding to obtain a larger shell very close to a sphere (Figure 2b). The two samples' detailed experimental parameters are shown in Table 1 and characters are shown in Table 2.

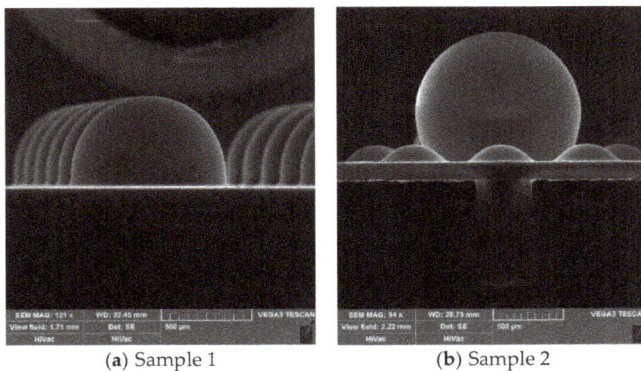

(a) Sample 1   (b) Sample 2

Figure 2. Scanning electron microscope (SEM) images of two different glass shells: (**a**) glass shell blown via an etched cavity without a foaming agent (H/D = 0.61); (**b**) glass shell blown via an etched cavity with quantified foaming agent (H/D = 0.80).

2. Height Model of a Micro Glass-Blown Shell

We have a height model to estimate the height of a glass-blown shell produced by an etched cavity with quantified foaming agent (Figure 3). The volume and shape of glass shell structures have been studied in [5], however, that study did not involve foaming agent. At a high temperature above the softening point of glass, the gas in the sealed cavity also includes the gas produced by foaming agent previously added in the etched cavity, this gas expands and increases the inner gas pressure before driving the high-temperature molten glass membrane to deform into a hollow spherical shell via the same surface pressure distribution. After 2–5 min the glass shell samples were removed rapidly to cool down in air to avoid the collapse of the glass shells. The shell shape was determined by the radius and depth of the etched cavity, the thickness of the glass wafer, the temperature at which the cavity was sealed, the temperature at which the glassblowing was executed, the mass of the foaming agent in sealed cavity, and even the cooling down process of the softened glass shell all have a significant influence on the final shape of the glass shell.

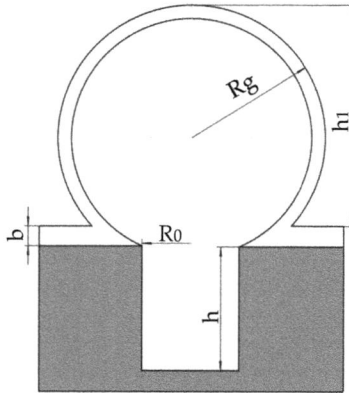

Figure 3. Height model of a micro glass-blown shell.

In the height model, we ignore the influence of gravity and the viscous force of the softened glass and assume the thickness of the glass shell is uniform; thus, the volume of inflated gas enclosed obeys the ideal gas law

$$PV = nRT \tag{1}$$

where P is the inner pressure in the glass shell that is equal to atmosphere when the glass shell heated in furnace reaches balanced state; n is the number of moles, consisting of two components, one is the gas inside the etched cavity when bonding a glass wafer and the other is the gas released by the foaming agent; R is the Boltzmann constant; T is the temperature in furnace. Because the foaming agent provides extra molding gas, the equation can transform into

$$V_g = \left(\frac{T_f}{T_b} - 1 \right) V_E + V_F \tag{2}$$

where V_g is the volume of the glass shell, T_f is the heating temperature in the furnace, and T_b is the temperature at which the etched cavity was bonded to a glass wafer. V_E is the volume of the etched cylindrical cavity, which can be written as

$$V_E = \pi R_0^2 h \tag{3}$$

where h is the depth of etched cavity. Regarding V_F, which is the volume of gas released by foaming agent when the bonded wafer transferred in furnace set a high temperature of T_f, we use $CaCO_3$ as the blowing agent, the chemical decomposition of which is described as

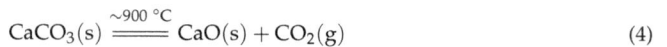

$$CaCO_3(s) \xrightleftharpoons{\sim 900\ °C} CaO(s) + CO_2(g) \tag{4}$$

The means to transfer the foaming agent will be explained in later section. We can control V_F by changing the number of moles of $CaCO_3$, n_0, which is equal to the number of moles of CO_2; thus, V_F can be derived as

$$V_F = \frac{n_0 R T_f}{P} \tag{5}$$

Because the shape of the glass shell is a spherical segment, V_g can be described by the mathematical formula

$$V_g = \pi h_1^2 \left(\frac{3R_g - h_1}{3} \right) \tag{6}$$

where h_1 is the height of the glass shell, R_g is the radius of the spherical segment, we assume the bottom radius of the glass shell is equal to the radius of etched cavity R_0, and V_g can also be described as another mathematical formula

$$V_g = \pi h_1 \left(\frac{3R_0^2 + h_1^2}{6} \right) \tag{7}$$

By combining (6) and (7), the mathematical equation describing the relationship between h_1 and R_g is

$$R_g = \frac{R_0^2 + h_1^2}{2h_1} \tag{8}$$

Thus, combining (6) and (8), the height of glass shell can be derived as

$$h_1 = \frac{\left[(3V_g + \sqrt{R_0^6 \pi^2 + 9V_g^2})\pi^2 \right]^{2/3} - R_0^2 \pi^2}{\pi \left[(3V_g + \sqrt{R_0^6 \pi^2 + 9V_g^2})\pi^2 \right]^{1/3}} \tag{9}$$

where $V_g = \left(\frac{T_f}{T_b} - 1 \right) V_E + \frac{n_0 R T_f}{P}$.

3. Materials and Methods

The fabrication process flow of a glass-blown spherical shell is shown in Figure 4. First a 200-nm thick layer of aluminum is sputtered onto a 1000-μm thick silicon wafer, and then the wafer is patterned using another layer of photoresist on the sputtered layer. Next, the aluminum mask is patterned by using aluminum etching liquid and removing the photoresist by acetone, followed by etching cylindrical cavities with 300 μm in radius and 800 μm in depth by DRIE. After this step, the aluminum layer is removed by aluminum etching liquid to obtain the substrate wafer with etched cavities.

For the process of adding the foaming agent, we choose $CaCO_3$ as the foaming agent because it meets the requirement of the subsequent process, with the thermal decomposition temperature (825 °C) that is lower than the temperature in the furnace (~900 °C) and higher than the temperature of anodic bonding (~400 °C). As a result, all the gas produced by $CaCO_3$ can contribute to the foaming process of the glass shell during the glass blowing process. However, transferring the solid foaming agent directly is too difficult to achieve. There are two main reasons for this difficulty: one is that there is no effective method or tool to place the solid $CaCO_3$ in a cavity with 300 μm radius, and the solid $CaCO_3$ (especially the powders) would result in bonding surface pollution that is difficult to remove; the other reason is that $CaCO_3$ is insoluble in most solvents. Thus, finding a proper means to place the foaming agent inside the micro cavity is the key problem to be solved.

Eventually, we utilized the precipitation reaction of Na_2CO_3 solution and $CaCl_2$ solution according to the chemical equation

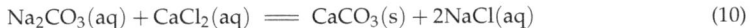

$$Na_2CO_3(aq) + CaCl_2(aq) \rightleftharpoons CaCO_3(s) + 2NaCl(aq) \tag{10}$$

With the reactor being the etched cavity, we use two syringe pumps (Pump 11 Elite, Harvard Apparatus, Holliston, MA, USA) and two microliter syringes (Shanghai Gaoge Industrial and Trading Company, Shanghai, China, model 1448348727, 10 μL). Moreover, we use a 34 G syringe needle (190 μm outer diameter and 60 μm inner diameter) to inject the solution because its outer diameter is smaller than the diameter of the cavity. We can control the quantity of $CaCO_3$ placed in an etched cavity by adjusting the injected volume and concentration of the two reacting solutions.

After the micro-injection process, the water is removed by drying on a hot plate, thus, the quantified amount of $CaCO_3$ remains in the cavity. Next, the silicon wafer is anodically bonded to a thin piece of Bf 33 glass (100 μm in thickness, Schott AG, Mainz, Germany) to isolate the atmosphere in the single cavity.

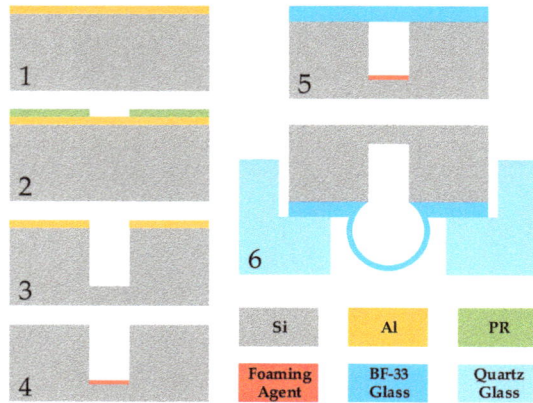

Figure 4. Fabrication process flow of a glass-blown spherical shell resonator.

In the step of glass blowing, we invert the bonded wafer by placing the wafer upside down onto a quartz glass stencil and then transferring the assembly together into a quartz tube furnace set to a high temperature (850 °C to 950 °C); the silicon wafer and the stencil are still stable, and the foaming agent would release the gas while the glass wafer would become softened and the increased gas pressure in the cavity would force the circular thin glass film on top of the etched cavity to become a spherical glass shell. The force of gravity of the softened glass is beneficial to obtain a larger glass shell (higher H/D). After an adequate heating time (120 s to 300 s), we take the glass shell samples out of the furnace and allow the glass shell to cool down rapidly to retain its shape, because if the heating time is too short, there's no enough time for glass to become soften and blown into hemisphere shape, and if the heating time is too long, the glass shell would deform even break because the softened glass would mostly flow to the shell vertex on the influence of gravity. The required blow-up time has been discussed in [5], we also perform series of contrast tests to determine the glass blowing characteristics.

4. Results and Discussion

4.1. Experimental Characteristics

In the experimental phase, researchers are interested in the effect of foaming agent on the shape of glass-blown shell; thus, contrast experiments were designed and executed. There are some differences between two experiments, with the most important distinction being whether the foaming agent was added into the etched cavity or not. We compare and analyze two glass shell samples we made: sample 1, which is shown in Figure 2a, was blown by a cavity without foaming agent; sample 2, which is shown in Figure 2b, was blown by a cavity that has 1.415 µg $CaCO_3$ in a sealed cavity. The other parameters, e.g., the radius of cavity R_0, the depth of cavity h, the thickness of glass b, the bonding temperature T_b, the temperature in the furnace for glass blowing T_f, and the heating time t, are all shown in Table 1.

Table 1. The experimental parameters of the two samples.

Parameter	R_0 (µm)	h (µm)	b (µm)	T_b (°C)	T_f (°C)	t (s)
Sample 1	300	800	129.82	360	900	180
Sample 2	300	800	132.85	400	900	180

The experiment results of the two samples are shown in Table 2. $V_{g\,predicted}$ is calculated by Equation (2), and $V_{g\,real}$ is calculated by Equation (6). The real volume of glass shell 2 is 276% more than the volume of shell 1, and the H/D also increases from 0.61 to 0.80, because the $CaCO_3$ can release extra foaming gas during the glass blowing process. Thus, we testify that the application of chemical foaming method in fabrication of a micro-hemisphere shell resonator can help researchers achieve a larger glass shell (higher volume shell or greater height). Furthermore, we can produce a micro glass-shell with a profile closer to a sphere compared with those micro-shells blown only by the sealed atmosphere.

Table 2. The experimental results of the two samples.

Parameter	$V_{g\,predicted}$ (nL)	$V_{g\,real}$ (nL)	$h_{1\,predicted}$ (µm)	$h_{1\,real}$ (µm)	$R_{g\,real}$ (µm)	H/D
Sample 1	192.92	207.69	592.85	512.12	422.78	0.61
Sample 2	1530.10	781.04	1400	950.19	592.09	0.80

4.2. Four-Node Wineglass Resonant Frequencies Simulation

Most HRGs utilize the four-node wineglass resonant mode ($n = 2$) of the HSR to sense the change of rotation, this mode is an important and inherent characteristic that varies from dozens of kHz to several MHz for different structures, materials, and processes.

Through the experimental process, we ultimately obtained hundreds of glass HSR samples, most of which have different H/D because of the different process parameters, and we confirmed that the most effective approach to increase the H/D is the chemical foaming method. In addition, we divided the HSR samples along the symmetry axis into two identical parts, after a detailed measurement of the HSR sectional dimensions via SEM. To study the influence of different shapes on the natural frequencies, we restructured five 3D models in SolidWorks (2010, Dassault Systèmes, Vélizy-Villacoublay, France) and then via COMSOL Multiphysics (5.2a, COMSOL Inc., Stockholm, Sweden) to simulate corresponding resonant frequencies of the four-node wineglass resonant mode. Figure 5 shows the simulated two degenerate four-node wineglass resonant mode shapes of an HSR, the phase difference between the two degenerate four-node wineglass resonant mode shapes is 45°.

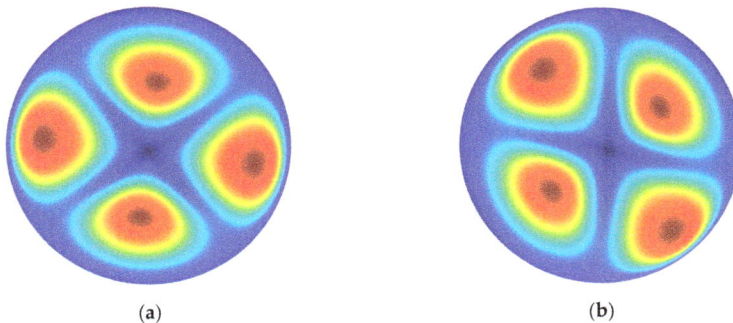

(a)   (b)

Figure 5. The simulated two degenerate four-node wineglass resonant mode ($n = 2$) shapes of an HSR: (**a**) The four-node wineglass resonant mode shape; (**b**) The degenerate four-node wineglass resonant mode shape.

The simulation results of the four-node wineglass resonant frequencies are shown in Table 3. The five samples were fabricated by using different process parameters, sample 6 and sample 7 were fabricated with the help of $CaCO_3$, other parameters are same, the thickness of glass wafer b is 100 µm and the depth of etched cavity h is 800 µm. HSRs with different profiles have different four-node wineglass resonant frequencies f_w ($n = 2$), we found that the f_w ($n = 2$) decrease from 2.08 MHz to 0.94 MHz through the addition of $CaCO_3$, it is a 54.8% decrease, enabling the HSR more sensitive to change of rotation, and it is convenient to set larger surrounding capacitive electrodes, allowing the HSR to be easier to drive and sense, showing the efficiency of the CFP method in fabrication of a micro glass HSR.

Table 3. The simulation results of four-node wineglass resonant frequencies.

Parameter	m_{CaCO_3} (μg)	R_0 (μm)	T_b (°C)	T_f (°C)	t (s)	h_1 (μm)	R_g (μm)	H/D	f_w (MHz)
Sample 3	none	300	400	950	210	324.41	452.26	0.36	2.08
Sample 4	none	300	360	890	180	505.03	434.35	0.58	2.00
Sample 5	none	300	360	930	180	501.84	439.48	0.57	1.88
Sample 6	1.415	300	400	900	180	989.99	633.14	0.78	1.11
Sample 7	1.887	500	400	900	180	1039.45	759.33	0.68	0.94

4.3. Glass HSR Blown by 200 μm Deep Cavity

Furthermore, we used a 500 μm silicon wafer to replace the 1000 μm silicon wafer, and we just etched 200 μm deep cavities instead of 800 μm deep cavities, the use of shallower cavities has many advantages. For example, the cost of DRIE is very high, so we can save approximately 75% of the expense of DRIE. In addition, we do not require the metal mask to resist such deep etching. Moreover, a thinner silicon wafer is beneficial for decreasing the final thickness of the HRG, which would help in the processes of DRIE, dicing, and anodic bonding, because 500 μm silicon wafers are more commonly used in micro-machining.

We control the mass of $CaCO_3$ in the 200 μm deep etched cavity while using the same values for the other process parameters, the SEM images of the glass shell samples are shown in Figure 6, and the corresponding results are shown in Table 4.

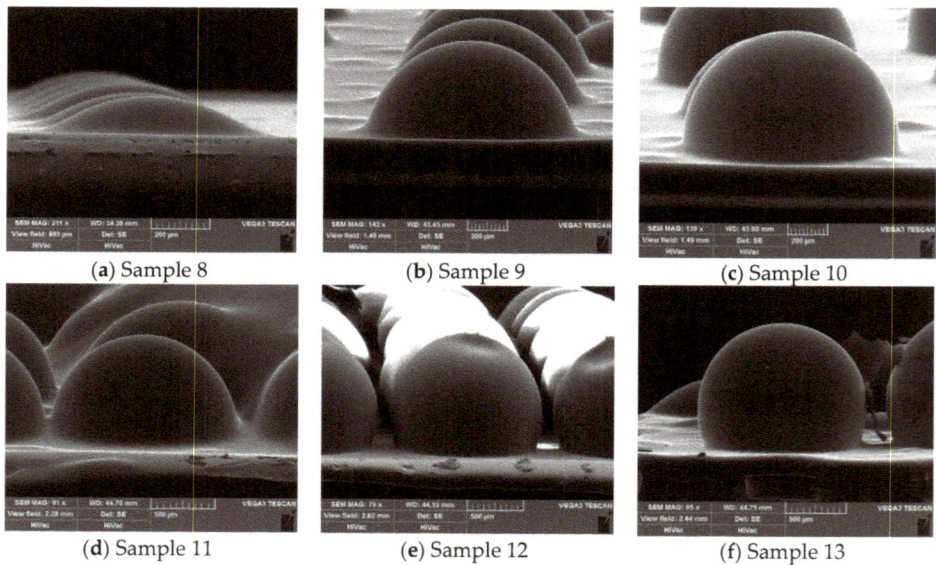

(**a**) Sample 8   (**b**) Sample 9   (**c**) Sample 10

(**d**) Sample 11   (**e**) Sample 12   (**f**) Sample 13

Figure 6. SEM images of the glass shell samples blown by 200 μm deep etched cavities.

Table 4. The experimental results of the glass shell samples blown by 200 μm deep etched cavities [1].

Parameter	m_{CaCO_3} (μg)	$V_{g\ predicted}$ (nL)	$V_{g\ real}$ (nL)	$h_{1\ predicted}$ (μm)	$h_{1\ real}$ (μm)	$D_{g\ real}$ (μm)	H/D	f_w (MHz)
Sample 8	none	43.683	27.93	250.66	133.39	1088.34	0.12	7.24
Sample 9	0.236	274.73	285.81	695.78	516.26	1026.86	0.50	1.59
Sample 10	0.472	505.78	446.36	897.76	682.42	1065.13	0.64	1.46
Sample 11	0.943	966.89	920.78	1200	823.83	1412.92	0.58	1.12
Sample 12	1.415	1429	1228.41	1300	1134.62	1363.88	0.83	1.05
Sample 13	1.887	1891.1	1437.96	1500	1205.46	1433.61	0.84	0.98

[1] These samples have same process parameters except for the mass of $CaCO_3$ in the 200 μm deep etched cavity, R_0 = 300 μm, *h* = 200 μm, T_b = 400 °C, T_f = 920 °C, *t* = 180 s.

By comparing the calculated volume and the real volume of the HSR samples in Figure 7a, we find that the real HSR volume increases when more $CaCO_3$ is added into the etched cavity. However, the increase slope decreases. For example, the real volume of shell sample 13 (which was blown by a 200 μm deep etched cavity with 1.887 μg of $CaCO_3$) is 51.48 times that of sample 8 (which was formed without the $CaCO_3$), although it should be 67.71 times greater in theory. The reason for this discrepancy may be the larger HSR shell has a thinner glass wall and more surface area, when we take the shell sample out of the furnace, the inner atmosphere of the HSR shell cool down more easily, which could result in the glass HSR shell shrink before the glass solidifies.

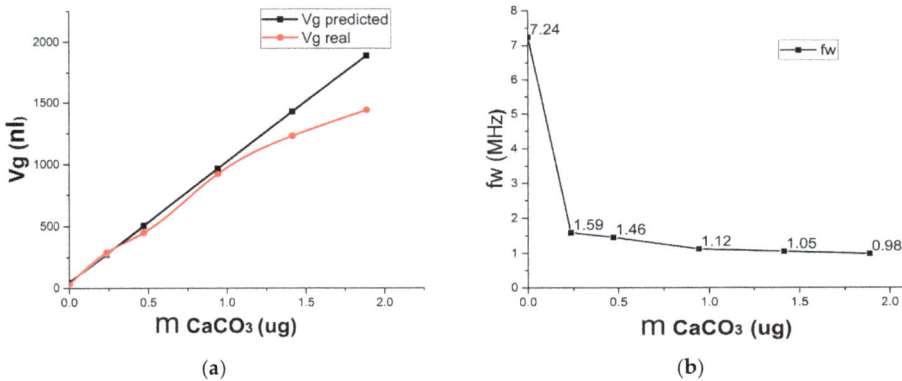

Figure 7. The effect of $CaCO_3$ on the volume and four-node wineglass resonant frequencies of HSR samples: (**a**) Calculated volume versus real volume of HSR samples; (**b**) The four-node wineglass resonant frequencies of the HSR samples.

The four-node wineglass resonant frequencies of HSR samples are shown in Figure 7b. Sample 8, which was blown by an etched cavity without $CaCO_3$, has a four-node wineglass resonant frequency at 7.24 MHz, and sample 9, which is blown by etched cavity with 0.236 μg $CaCO_3$ that practically provides an extra 9.23 times volume as that of sample 8, has a four-node wineglass resonant frequency at 1.59 MHz, which is 78% lower than that of the control group of sample 8 (7.24 MHz). The reason for the decrease of f_w (*n* = 2) is that the larger HSR has a thinner glass wall and lower stiffness. Note that the f_w (*n* = 2) decrease slope of other HSR samples is lower than that of samples 8 and 9, showing the efficiency of CFP method in the fabrication of the HSR. The H/D value also increases from 0.12 to 0.50 with the addition of $CaCO_3$, and the profile of sample 9 is a half sphere. Regarding samples 10, 11, and 13, their H/D values all exceed 0.5, and the largest one, sample 13, has a 4/5 sphere profile. We also attempted other approaches, such as lowering the sealed temperature or increasing the blowing temperature; the improvement of the HSR volume was minor, and those approaches are limited by the process conditions, so the CFP method is a better choice when we want to get a larger glass HSR close to a sphere.

We conclude that we could obtain a larger HSR via application of the chemical foaming method, which provides extra foaming gas. In addition, the volume of the HSR blown by the 200 μm deep etched cavity with the quantified addition of foaming agent can approach (even exceed) the HSR blown by the 800 μm deep etched cavity without foaming agent. The proposed approach not only reduces process time and cost but also presents a means to reduce the dimensions of the HRG. Furthermore, the larger HSR has several advantages, for example, greater surface area is helpful for setting larger surrounding drive and sense capacitive electrodes, which can increase the HSR detection sensitivity of rotation change.

Although the efficiency and economy of the introduction of CFP have been proven in fabrication and experiment tests of the HSR, several difficulties remain need to be solved. For example, injecting two reaction solutions must be very carefully performed; thus, we should ensure the liquid does not overflow onto the surface of the silicon wafer to avoid the failure of anodic bonding. In addition, the volumes of the two reacting solutions injected are very small (25 nL to 200 nL) and must be accurate; thus, the proposed approach has a high requirement of location accuracy and injection control precision for the platform used to add the foaming agent. Moreover, there are many factors that can influence the final shape of the glass HSR, and previous processes have absolutely essential influences on the final glass blowing result; thus, exploring a standard and successful process flow is of great importance.

5. Conclusions

A novel approach involving the application of the chemical foaming method in the fabrication of micro-hemispherical shell resonators was developed to increase the volume of the glass-blown shell, achieving a 51.48-fold increase compared to the traditional method without the foaming agent. The larger HSR with greater surface area is helpful for setting larger surrounding drive and sense capacitive electrodes, which can increase the HSR detection sensitivity of rotation. A height model was first presented to estimate the shape of the glass-blown shell with chemical foaming agent, and comparison experiments were performed to prove the efficiency of the new approach. We also developed a method to miniaturize the dimension of the glass HSR; this method can reduce the cost and increase the sensitivity of a glass HSR. This method may inspire related research in the design and fabrication of HRG.

Acknowledgments: The research was financially supported by the National Natural Science Foundation of China (51405388, 51775447).

Author Contributions: Jianbing Xie conceived and guided the research; Lei Chen conceived and designed the experiments; Lei Chen and Hui Xie performed the experiments; Lei Chen and Guangcheng Liu analyzed the data; Jinqiu Zhou contributed the analysis of height model; Lei Chen wrote this paper.

Conflicts of Interest: The authors declare no conflict of interest.

References

1. Matthews, A.D.; Bauer, D.A. The hemispherical resonator gyro for precision pointing applications. In Proceedings of the SPIE 2466, Space Guidance, Control, and Tracking II, Orlando, FL, USA, 12 June 1995; pp. 128–139.
2. Prikhodko, I.P.; Zotov, S.A.; Trusov, A.A.; Shkel, A.M. Microscale glass-blown three-dimensional spherical shell resonators. *J. Microelectromech. Syst.* **2011**, *20*, 691–701. [CrossRef]
3. Pai, P.; Chowdhury, F.K.; Pourzand, H.; Tabib-Azar, M. Fabrication and testing of hemispherical MEMS wineglass resonators. In Proceedings of the IEEE 26th International Conference on Micro Electro Mechanical Systems (MEMS), Taipei, Taiwan, 20–24 January 2013; pp. 677–680.
4. Cho, J.; Woo, J.K.; Yan, J.; Peterson, R.L.; Najafi, K. A high-Q birdbath resonator gyroscope (BRG). In Proceedings of the Transducers 2013, Barcelona, Spain, 16–20 June 2013; pp. 1847–1850.

5. Eklund, E.J.; Shkel, A.M. Glass blowing on a wafer level. *J. Microelectromech. Syst.* **2007**, *16*, 232–239. [CrossRef]

6. Giner, J.; Gray, J.M.; Gertsch, J.; Bright, V.M.; Shkel, A.M. Design, fabrication, and characterization of a micromachined glass-blown spherical resonator with in-situ integrated silicon electrodes and ALD tungsten interior coating. In Proceedings of the IEEE MEMS 2015, Estoril, Portugal, 18–22 January 2015; pp. 805–808.

7. Sorenson, L.D.; Gao, X.; Ayazi, F. 3-D micromachined hemispherical shell resonators with integrated capacitive transducers. In Proceedings of the MEMS 2012, Paris, France, 29 January–2 February 2012; pp. 168–171. Available online: http://ieeexplore.ieee.org/document/6170120/ (accessed on 31 October 2017). [CrossRef]

8. Bernstein, J.J.; Bancu, M.G.; Bauer, M.J.; Cook, E.H.; Kumar, P.; Newton, E.; Nyinjee, T.; Perlin, G.E.; Ricker, J.A.; Teynor, W.A.; et al. High Q diamond hemispherical resonators: Fabrication and energy loss mechanisms. *J. Micromech. Microeng.* **2015**, *25*, 085006. [CrossRef]

9. Heidari, A.; Chan, M.L.; Yang, H.A.; Jaramillo, G.; Taheri-Tehrani, P.; Fonda, P.; Najar, H.; Yamazaki, K.; Lin, L.; Horsley, D.A. Hemispherical wineglass resonators fabricated from the microcrystalline diamond. *J. Micromech. Microeng.* **2013**, *23*, 125016. [CrossRef]

10. Shao, P.; Tavassoli, V.; Mayberry, C.L.; Ayazi, F. A 3D-HARPSS polysilicon microhemispherical shell resonating gyroscope: Design, fabrication, and characterization. *IEEE Sens. J.* **2015**, *15*, 4974–4985. [CrossRef]

11. Cho, J.; Nagourney, T.; Darvishian, A.; Shiari, B.; Woo, J.K.; Najafi, K. Fused silica micro birdbath shell resonators with 1.2 million Q and 43 second decay time constant. In Proceedings of the Solid-State Sensors, Actuators and Microsystems Workshop, Hilton Head Island, SC, USA, 8–12 June 2014; pp. 103–104. Available online: http://wims.eecs.umich.edu/publications/papers/fused-silica-micro_j-cho.pdf (accessed on 31 October 2017).

12. Senkal, D.; Ahamed, M.J.; Trusov, A.A.; Shkel, A.M. High temperature micro-glassblowing process demonstrated on fused quartz and ULE TSG. *Sens. Actuators A Phys.* **2012**, *201*, 525–531. [CrossRef]

13. Zotov, S.A.; Trusov, A.A.; Shkel, A.M. Three-dimensional spherical shell resonator gyroscope fabricated using wafer-scale glassblowing. *J. Microelectromech. Syst.* **2012**, *21*, 509–510. [CrossRef]

14. Senkal, D.; Prikhodko, I.P.; Trusov, A.A.; Shkel, A.M. Micromachined 3-D glass-blown wineglass structures for vibratory MEMS applications. In Proceedings of the Technologies for Future Micro-Nano Manufacturing, Napa, CA, USA, 8–10 August 2011; pp. 166–169.

15. Shang, J.T.; Chen, B.Y.; Lin, W.; Wong, C.P.; Zhang, D.; Xu, C.; Liu, J.W.; Huang, Q.A. Preparation of wafer-level glass cavities by a low-cost chemical foaming process (CFP). *Lab Chip* **2011**, *11*, 1532–1540. [CrossRef] [PubMed]

16. Luo, B.; Shang, J.T.; Zhang, Y.Z. Hemipherical wineglass shells fabricated by a chemical foaming process. In Proceedings of the 16th International Conference on Electronic Packaging Technology, Changsha, China, 11–14 August 2015; pp. 951–954.

micromachines

MDPI

Article

Micro-Electromechanical Acoustic Resonator Coated with Polyethyleneimine Nanofibers for the Detection of Formaldehyde Vapor

Da Chen [1,2,*], Lei Yang [2], Wenhua Yu [2], Maozeng Wu [2], Wei Wang [2] and Hongfei Wang [2]

[1] State Key Laboratory of Mining Disaster Prevention and Control Co-founded by Shandong Province and the Ministry of Science and Technology, Shandong University of Science and Technology, Qingdao 266590, China

[2] College of Electronics, Communications, and Physics, Shandong University of Science and Technology, Qingdao 266590, China; ylei1994@163.com (L.Y.); 17860756595@163.com (W.Y.); wumaozeng1993@163.com (M.W.); skdwangwei1zqf@163.com (W.W.); phywjj@163.com (H.W.)

* Correspondence: chenda@sdust.edu.cn; Tel.: +86-0532-8605-7555

Received: 19 December 2017; Accepted: 26 January 2018; Published: 1 February 2018

Abstract: We demonstrate a promising strategy to combine the micro-electromechanical film bulk acoustic resonator and the nanostructured sensitive fibers for the detection of low-concentration formaldehyde vapor. The polyethyleneimine nanofibers were directly deposited on the resonator surface by a simple electrospinning method. The film bulk acoustic resonator working at 4.4 GHz acted as a sensitive mass loading platform and the three-dimensional structure of nanofibers provided a large specific surface area for vapor adsorption and diffusion. The ultra-small mass change induced by the absorption of formaldehyde molecules onto the amine groups in polyethyleneimine was detected by measuring the frequency downshift of the film bulk acoustic resonator. The proposed sensor exhibits a fast, reversible and linear response towards formaldehyde vapor with an excellent selectivity. The gas sensitivity and the detection limit were 1.216 kHz/ppb and 37 ppb, respectively. The study offers a great potential for developing sensitive, fast-response and portable sensors for the detection of indoor air pollutions.

Keywords: film bulk acoustic resonator; formaldehyde; gas sensor; nanofibers

1. Introduction

Formaldehyde, usually derived from household materials, is one of the most common indoor air pollutants. There is a strong demand for a sensitive, fast-response and portable method to detect formaldehyde for indoor environmental monitoring due to its high carcinogenicity [1]. The traditional spectroscopy, chromatography and mass spectrometry are very hypersensitive, accurate and reliable, but they are limited by the large-scale equipment, professional operation and unable to detect formaldehyde at a customer's home [2]. The solution leads to the development of smart formaldehyde sensors with small device size and rapid response speed. So far, metal oxide semiconductor (MOS) [3–5], carbon nanotubes (CNTs) [6–8], conductive polymer [9] have been used to fabricate formaldehyde sensors based on field effect, resistive and electroacoustic principles. Over the past decade, the technical progress in micro-electromechanical systems (MEMS) brings a novel development direction for the microsensors.

Film bulk acoustic resonator (FBAR) is a promising microelectromechanical system (MEMS) resonator and has obtained the preliminary success in radio frequency communication technologies [10–12]. Moreover, its applications for gas [13–16] and biochemical detections [17–20] have received attention thanks to the high sensitivity and micron-scale size. Compared with the conventional electroacoustic resonator such as quartz crystal microbalance (QCM), the important advance of FBAR is the use of

1–2 microns-thick piezoelectric films to replace the crystal plates, which provides a fundamental working frequency at several gigahertz and enough mass sensitivity to probe a single gas molecules layer [21,22]. In addition, FBAR is fabricated by standard MEMS process, thereby realizing the ability to inexpensively combine a number of sensors on a chip and integrate them with the analytical circuits. For gas-sensing applications, the FBAR usually works as a mass-loading platform. A sensitive layer is coated on the device surface to absorb the target molecules. The small additional mass on the sensing layer is detected by monitoring the variation of resonant frequency. As a result, the properties of sensitive coating determine the molecule recognitions, and directly affect the sensitivity, stability and reversibility of the sensor. Up to now, a variety of sensitive coatings, such as polymers [23,24], proteins [25–27], aptamers [28–31], enzymes [32,33], supramolecular monolayers [34,35], hydrophilic film [23] and CNTs [36] have been employed for FBAR sensors to achieve the selectivity for different analytes.

Polyethyleneimine (PEI) and its derivatives are regarded as an appropriate formaldehyde-sensitive material since PEI can efficiently adsorb formaldehyde molecules via a reversible reaction of primary amines. Therefore, polystyrene (PS)/PEI [37], poly(vinyl alcohol) (PVA)/PEI [38], TiO_2/PEI [39], and CNTs/PEI composites [40] have been reported as the sensitive coating of mass-loading sensors for the detection of formaldehyde vapor. On the other hand, in the view of the high sensitivity of FBAR devices, the nanostructured materials featured with high surface areas and numerous sites are the potential coating for gas sensing.

In this paper, we developed a promising strategy to combine PEI nanofibers and the FBAR with high mass sensitivity to construct a formaldehyde microsensor. The PEI nanofibers were directly deposited on an AlN FBAR surface by a simple electrospinning method. Benefiting from the high working frequency at 4.4 GHz, the proposed FBAR was able to measure the ultra-small mass change produced by the interaction between formaldehyde molecules and the PEI nanofibers with linear response characteristics, fast response/recovery rate and excellent selectivity.

2. Device Configuration and Fabrication

2.1. Schematic Structure and Sensing Mechanism

Figure 1a,b shows the schematic structure, sensing mechanism and the photomicrograph of the FBAR formaldehyde sensor. The major structure of FBAR (300×150 μm^2) is a sandwiched Au (100 nm)/AlN (1 μm)/Mo (100 nm) piezoelectric stack built on a Si_3N_4 sputtered layer (0.6 μm). The PEI nanofibers were deposited on the surface of top Au electrode as the specific sensitive coating. The active resonance and sensing area are overlapped between the two electrodes (3296 μm^2). When the FBAR sensor is exposed to gaseous formaldehyde, the reversible nucleophilic addition reaction happens between the amines of PEI and the vapor molecules at room temperature [41]. In fact, there is a σ-bond and a π-bond in the formaldehyde molecule. Because of the difference in electron affinity, the oxygen atom side shows electronegativity while the positive side of carbon atoms can be considered as the electrophile. In the amine group, the lone pair in the nitrogen atoms works as the nucleophile and engages in the reaction with the π-bond in formaldehyde molecules. Therefore, the vapor absorption can be measured by monitoring the downshift of resonant frequency based on the mass-sensitive mechanism.

In order to minimize the environmental disturbance, the practical differential frequency method was used to extract the sensing response as shown in Figure 1c. For this purpose, both the FBAR coated with PEI nanofibers and the reference FBAR device (without coating) were wire-bonded to a printed circular board (PCB) and packaged in the test chamber (Figure 1d). Upon the exposure of formaldehyde, the resonant frequency of the former device decreased while the reference frequency measured from the latter kept stable. The frequency difference between the two devices was read out by a testing circuit as the sensing response to formaldehyde vapor. All components in the test circuit were off-the-shelf. At first, the two FBAR devices were driven by independent Colpitts oscillators.

The frequency signals were put into the mixer and then passed through the balun transformer, low-pass filtered, the amplified, waveform converter and frequency dividing circuits. At last, a micro controller unit was used to count, read out and store the differential frequency. The details of the circuit design were shown in Supplementary Material (Figure S1).

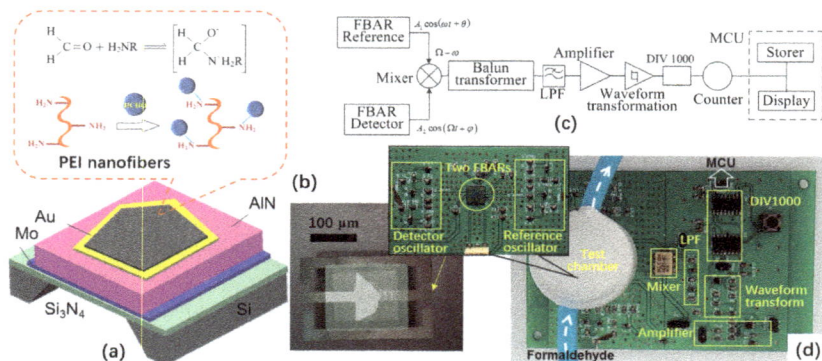

Figure 1. (**a**) Schematic structure and sensing mechanism of the film bulk acoustic resonator (FBAR) sensor coated with polyethyleneimine (PEI) nanofibers; (**b**) Photomicrograph of the fabricated device; (**c**) Block diagram of the testing system based on differential frequency processing; (**d**) Completed circuit board of the testing system.

2.2. Fabrication of Film Bulk Acoustic Resonator (FBAR) Device

The FBAR sensor was fabricated with a four-mask process using standard MEMS technology as shown in Figure 2a. First, a low-stress Si_3N_4 layer was grown on both sides of a (100) silicon wafer by low-pressure chemical vapor deposition. Then, an initial cavity was formed on one side of the wafer by wet etching (80 °C KOH) with the patterned Si_3N_4 layer as the mask. About 15-μm-thick silicon was left to support the following process on the other side of the wafer. Next, the Mo/AlN/Au piezoelectric stack was prepared by radio frequency magnetron sputtering and patterned by conventional photolithography technique. Finally, the residual silicon under the stack was etched by deep reactive ion etching to isolate the resonator acoustically from the substrate.

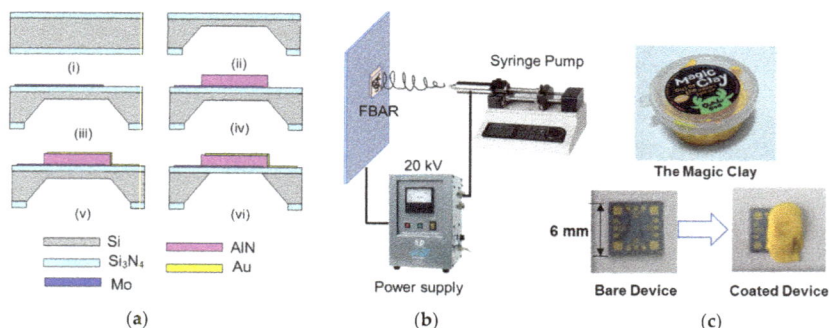

Figure 2. (**a**) Fabrication process of the FBAR formaldehyde sensor: (i) Growth of Si3N4 film; (ii) Etching of one side of the silicon wafer; (iii) Deposition and pattern of bottom Mo electrode; (iv) Deposition of AlN film; (v) Preparation of top Au electrode; (vi) Dry etching of the residual silicon. (**b**) Schematic diagram illustrating the electrospinning deposition of PEI nanofibers. (**c**) Photographs of the FBAR devices before and after the coating of clay.

2.3. Electrospinning Deposition of Polyethyleneimine (PEI) Nanofibers

After the fabrication process, the silicon wafer was cut into small pieces with the area of 6 mm ×
6 mm (as shown in Figure 2c). Two FBAR devices on the opposite side were used in this experiment
as the detector and reference device, respectively. The reference device without sensitive layer was
manually coated with kid's magic clay (DoDoLu, Zhigao colored clay Co., Ltd., Jinhua, China) before
the electrospinning process (see Figure 2c). The main components of magic clay are polyethylene
ethanol, cross-linking agent and water. The viscous magic clay was attached to the device surface and
then was easily removed after natural drying without any damage on the device structure. The PEI
aqueous solution (10 wt %, MW = 25,000, Macklin Biochemical Co., Ltd., Shanghai, China) was used
as the electrospinning solution. The feed rate of the solutions was regulated at 5 mL/h by a syringe
pump (LSP02, Longer Precision Pump Co., Ltd., Baoding, China). By applying the voltage of 20 kV
between the syringe and the conductive sheet at a tip-to-collector distance of 20 cm, the nanofibers
were continuously deposited on the FBAR device surface. After electrospinning, the device was dried
at 40 °C in vacuum for 30 min to evaporate the solvent. In addition, a flat PEI film was spin-coated
(30 wt % PEI solution, 2500 rpm for 45 s) on the surface of another FBAR device to compare the
formaldehyde sensing characteristics.

3. Results and Discussion

3.1. Characterization of the Resonator

The surface morphology of the sensitive coating on the FBAR surface was observed by a field
emission scanning electron microscope (FE-SEM, S-4800 Hitachi, Tokyo, Japan) as shown in Figure 3.
The average diameter of the PEI fibers was about 40 nm with random orientations. The distribution
became denser with the increase of deposition time, which had an obvious effect on the formaldehyde
response. Compared with the layer imbedded with nanoparticles [39] or nanotube [40] prepared by
spin coating, the electrospinning method forms the three-dimensional structure, which can provide a
larger specific surface area for vapor adsorption and diffusion.

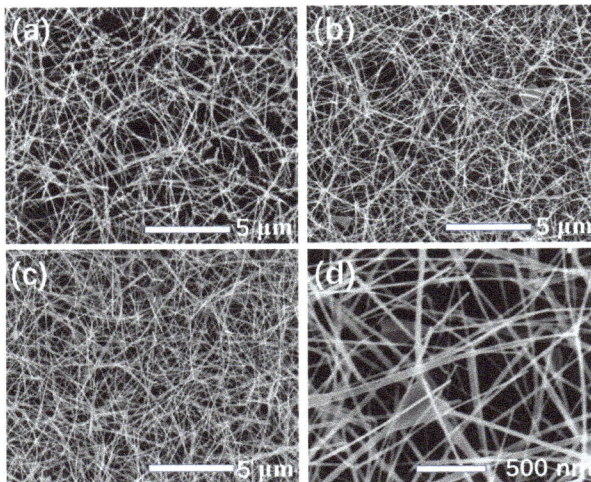

Figure 3. Typical FE-SEM images of the PEI nanofibers on the FBAR surface for the deposition times of
(**a**) 50 s, (**b**) 150 s and (**c**) 300 s. The detail view of the nanofibers is shown in (**d**).

The thickness of the PEI nanofiber layers was estimated by the cross-view SEM images as shown in Figure 4. The thickness of the PEI nanofiber layer exhibited a linearly dependence on the deposition time with the rate of 3.185 nm/s. Figure 5 shows the admittance of the FBAR sensors coating with different amounts of nanofibers. The device frequency linearly went down from 4464.15 MHz to 4402.41 MHz with increasing nanofibers (Figure 6a). In addition, the Q factors of the FBAR devices were determined by the Butterworth–Van Dyke model [42] to evaluate the energy loss because of the sensitive coating. As shown in Figure 6b, the coated PEI nanofibers only caused a small influence on Q factors (<5%) when the deposition time was less than 150 s. However, the loading of superfluous fibers resulted in a dramatic degradation of Q factors, which may be ascribed to the scattering and absorption of acoustic energy from the porous structure on the wave propagation path. Besides, the resonant frequency of the bare device shows a temperature coefficient of frequency (TCF) of −52.2 ppm/°C [43], which can be attributed to the thermal expansion of the piezoelectric film and the change of the acoustic velocity.

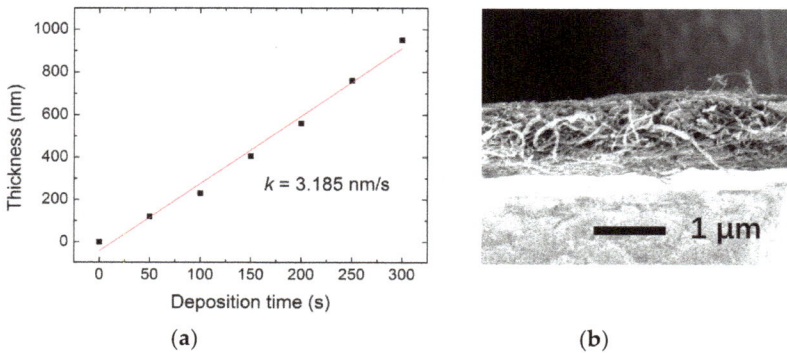

(a) **(b)**

Figure 4. (a) Dependence of the thickness of the PEI nanofiber layer on the deposition time; (b) Typical cross-view SEM image of the PEI nanofiber layer with the deposition times of 300 s.

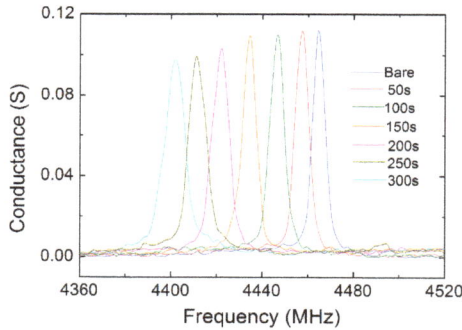

Figure 5. The admittance curves of the bare FBAR device and PEI nanofibers-coated FBAR devices. The times of electrospinning deposition were 50–300 s.

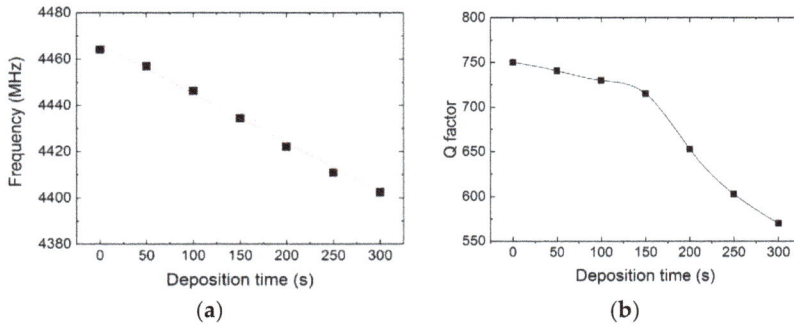

Figure 6. The changes of (**a**) resonant frequency and (**b**) Q factor of the PEI nanofibers-coated FBAR devices for different deposition times.

3.2. Effect of Deposition Time on the Formaldehyde Response

In the gas-sensing tests, the formaldehyde samples with required concentration were obtained by multiple dilution of the standard vapor (Changyuan Gas, Nanjing, China). All the gas tests were performed at the constant temperature condition (22 °C). Figure 7 shows the time-dependent frequency response of the FBAR sensors coated with flat PEI film and different amounts of PEI nanofibers in the formaldehyde vapor with the concentration of 300 ppb. Every sensing cycle comprised both absorption and desorption processes. At first, nitrogen was delivered to the test chamber and the stable resonant frequency was recorded as the baseline. After the formaldehyde was injected, the formaldehyde molecules were absorbed on the sensitive coating resulting in the downshift of resonant frequency. When the nitrogen was introduced again, the formaldehyde molecules were blown off from the sensitive coating and thus the response gradually raised to the baseline.

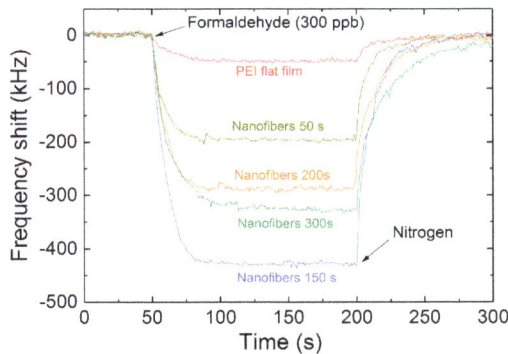

Figure 7. Time-dependent frequency response of the FBAR sensors coated with PEI flat film and different amounts of PEI nanofibers when exposed to 300 ppb formaldehyde vapor. All the tests were performed at room temperature and 40% relative humidity.

The structure and amount of the sensitive coating had an obvious influence on the absorption/ desorption behavior of formaldehyde. At the formaldehyde concentration of 300 ppb, the response of the flat film coated sensor was very small (~50 kHz). As expected, the fibrous structure could absorb more molecules and significantly enhance the sensitivity. The adsorption of formaldehyde onto the sensitive coating is maintained by the polymeric amines in PEI via the reversible nucleophilic addition reaction to form Schiff base [38]. As shown in the SEM images, the PEI fibers were connected to each other, forming net-like scaffolds, which increased the surface area and absorption sites and allowed the formaldehyde molecules to diffuse more easily inside the large space to react with the amine groups of PEI.

In particular, the sensor with the appropriate deposition time (150 s) exhibits the highest frequency downshift (~430 kHz) among the devices. Some references considered the coupled resonance between the microstructure coating and acoustic wave sensor [44–47]. However, in our study, the frequency shows a linearly negative shift with increasing of deposition time. The "drop and jump" change (the hallmark feature of coupled resonance) was not observed in Figure 6. On the other hand, the theoretical model proposed by Mansfeld [48] indicates that the mass-loading sensitivity increases sharply when the thickness of sensitive layer is close to a quarter of the acoustic wavelength. The Young modulus and acoustic velocity of solid-state PEI is 3.1–3.7 GPa and 2100–2300 m/s, respectively [49]. For the FBAR working frequency at 4.4 GHz, the corresponding quarter of the acoustic wavelength is 470~520 nm, which is very close to the thickness at the deposition time of 150 s in consideration of the velocity deviation between the solid state and nanofibers.

3.3. Sensitive Performance of the Optimized Sensor

On the basis of the above results, the optimized deposition time of 150 s was selected for further tests. Figure 8a shows the dynamic measurement of the optimized FBAR sensor exposed to the formaldehyde pulses with increasing concentrations. The fast adsorption and complete desorption took place with the response time of 10–25 s and the recovery time within 60 s. A saturated response was observed at vapor concentrations higher than 600 ppb, which is associated with the saturable adsorption of the PEI nanofibers. However, this detectable range could be accepted for the indoor air monitoring because serious throat and nasal irritation will occur when the formaldehyde concentration reaches about 1 ppm. As shown in Figure 8b, the frequency downshift of FBAR sensor was linearly proportional to the formaldehyde concentration before saturation with the regression coefficient (R^2) of 0.9921. The gas sensitivity, defined as the slope of calibration curve, was estimated to be -1.257 kHz/ppb over the linear region. The limit of detection (LOD) is given by LOD = $3\sigma/S$, where S is the gas sensitivity of the sensor and σ is the average noise of the intrinsic frequency [50], respectively. Benefiting from the high mass-sensitivity and the porous structure of PEI fibrous coating, the LOD of the FBAR sensor was as low as 37 ppb ($\sigma \approx 15$ kHz). In comparison, the formaldehyde detection limit of a recently reported QCM sensor is 600 ppb [40]. The sensor noise determines the minimum detectable frequency shift and thus is related to the detection limit. In this study, the average noise level was about 15 kHz at the resonant frequency of 4.4 GHz, which is higher than that of commercialized QCM systems (typically 0.01 Hz/8 MHz). Even so, the FBAR sensor still exhibits a very high sensitivity and low detection limit. It is believed that the sensitivity can be further improved by optimizing the device structure and test circuit.

For the mass-loading sensors, the higher working frequency produces a larger frequency shift with the same mass change. Therefore, the FBAR sensor exhibits a gas sensitivity that was about ten times higher than that of the QCM formaldehyde sensors [37–39]. The LOD of FBAR sensor reached the same levels of MOS [3–5] and CNTs sensors [6–8]. Remarkably, in our case, we are able to employ the operation at room temperature and the integration capability into micro-electromechanical systems. Furthermore, compared with the analog resistance or current signals produced by MOS and CNTs sensors, the digitized frequency response used by FBAR is more robust and can be read out directly without the need of analog–digital conversion. Benefiting from the micrometer-scale size and the MEMS fabrication, an e-nose could be constructed by integrating a large number of FBAR devices with different sensitive coatings. Each sensor is sensitive to some analytes with different responses; thus, a fingerprint pattern is generated for specific target recognition from the complex environment. For example, Yao el al. [34] demonstrated an e-nose type gas sensor for the selective detection of volatile organic compounds based on the FBAR sensor array functionalized with four supramolecular monolayers (p-tert-butyl calix[8]-arene, porphine, β-cyclodextrin, and cucurbit[8]uril.).

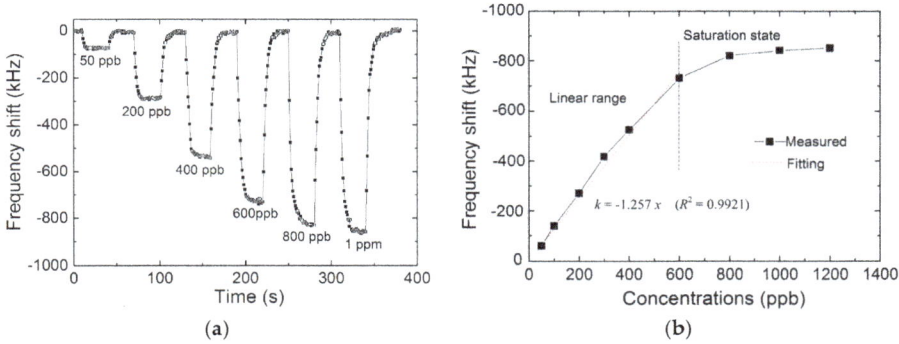

Figure 8. (**a**) Dynamic measurement of the FBAR sensor coated with PEI nanofibers exposed to the formaldehyde pulses with increasing concentrations at room temperature and 40% relative humidity; (**b**) Dependence of the frequency shift on the formaldehyde concentration.

3.4. Influence of Relative Humidity

In order to evaluate the influence of humidity on the sensing characteristics, the formaldehyde vapor mixed with saturated water vapor was delivered to the test chamber. Figure 9a shows the frequency response of the optimized FBAR sensor in three typical vapor concentrations measured at different ambient humidity. Both the frequency shift and the gas sensitivity were obviously enhanced with the increase of humidity. A possible reason is that the formaldehyde molecules could easily attach to the water molecules via hydrogen bonds [51]. In the humid environment, more water molecules were attached on the hydrophilic amine groups of PEI, and thus the adsorption capacity of the sensitive coating was improved.

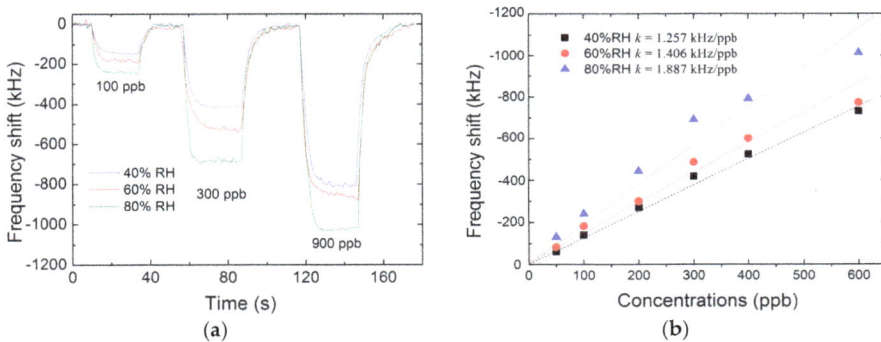

Figure 9. (**a**) Real-time frequency response of the optimized FBAR sensor exposed to three typical formaldehyde concentrations at the relative humidity of 40%, 60% and 80%. (**b**) Plots of frequency shift as a function of formaldehyde concentration at the relative humidity of 40%, 60% and 80%.

3.5. Selectivity of the Sensors

The FBAR sensor was tested against several potential indoor pollution vapors, including ethanol, acetone, benzene, dichloromethane, toluene and chloroform. As shown in Figure 10, the response to formaldehyde was about ten times higher than that of ethanol (the second largest response) indicating the excellent selectivity of the FBAR sensor coated with PEI nanofibers. As discussed before in Section 2.1, the main absorbing mechanism to formaldehyde is the nucleophilic addition reaction between the carbon atoms in π-bonds (electrophile) and the nitrogen atoms in amine groups (nucleophile). As a result, the specific attachment to formaldehyde is significantly stronger than the physical or nonspecific adsorption of the interference vapors. Similar results were also observed in other PEI-based sensors [36–39].

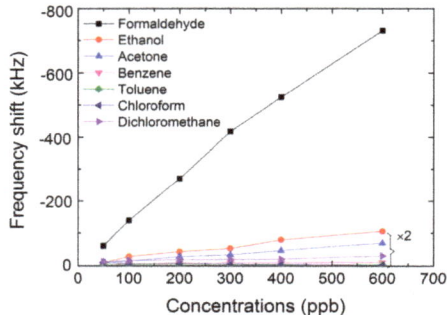

Figure 10. Plots of the frequency shift as a function of different concentration for formaldehyde and potential indoor pollution vapors.

4. Conclusions

In summary, we deposited PEI nanofibers on the FBAR mass-loading sensor by electrospinning and demonstrated its application for the detection of trace formaldehyde at room temperature. The interconnected PEI fibers formed a porous three-dimensional structure and provided a larger specific surface area to absorb formaldehyde molecules. The FBAR sensor exhibits a linear frequency downshift with increasing vapor concentration and excellent selectivity to formaldehyde with respect to other conventional organic vapors. The gas sensitivity is 1.216 kHz/ppb with the LOD of 37 ppb. The proposed FBAR device is a promising candidate as a mass-sensitive platform, and the studies of the electrospinning would benefit the developing of new sensitive coating for FBAR gas sensors.

Supplementary Materials: The following are available online at www.mdpi.com/2072-666X/9/2/62/s1, Figure S1: Ddetails of the circuit design of the film bulk acoustic resonator (FBAR) testing system.

Acknowledgments: This work was supported by the National Science Foundations of China (Grant No. 61574086, 11504207), National Science Foundation of Shandong Province (Grant No. ZR2014AQ006), Qingdao science and technology program of basic research projects (Grant No. 15-9-1-78-jch), Project of Shandong Province Higher Educational Science and Technology Program (Grant No. J15LN17) and the Tai'shan Scholar Engineering Construction Fund of Shandong Province of China.

Author Contributions: Da Chen proposed the main idea and designed the device structure. Lei Yang and Hongfei Wang fabricated the device and prepared the nanofibers. Wenhua Yu, Maozeng Wu and Wei Wang tested the resonant characterization and performed the gas-sensing experiments. Lei Yang and Da Chen wrote the paper.

Conflicts of Interest: The authors declare no conflict of interest.

References

1. Davis, M.E.; Blicharz, A.P.; Hart, J.E.; Laden, F.; Garshick, E.; Smith, T.J. Occupational exposure to volatile organic compounds and aldehydes in the U.S. trucking industry. *Environ. Sci. Technol.* **2007**, *41*, 7152–7158. [CrossRef] [PubMed]

2. Marć, M.; Zabiegała, B.; Namieśnik, J. Testing and sampling devices for monitoring volatile and semi-volatile organic compounds in indoor air. *TrAC Trends in Anal. Chem.* **2012**, *32*, 76–86. [CrossRef]

3. Park, H.J.; Kim, J.; Choi, N.J.; Song, H.; Lee, D.S. Nonstoichiometric Co-rich ZnCo$_2$O$_4$ hollow nanospheres for high performance formaldehyde detection at ppb levels. *ACS Appl. Mater. Interfaces* **2016**, *8*, 3233–3240. [CrossRef] [PubMed]

4. Hu, P.; Han, N.; Zhang, D.; Ho, J.C.; Chen, Y. Highly formaldehyde-sensitive, transition-metal doped ZnO nanorods prepared by plasma-enhanced chemical vapor deposition. *Sens. Actuators B* **2012**, *169*, 74–80. [CrossRef]

5. Zheng, Y.; Wang, J.; Yao, P. Formaldehyde sensing properties of electrospun NiO-doped SnO$_2$ nanofibers. *Sens. Actuators B* **2011**, *156*, 723–730. [CrossRef]

6. Yoosefian, M.; Raissi, H.; Mola, A. The hybrid of Pd and SWCNT (Pd loaded on SWCNT) as an efficient sensor for the formaldehyde molecule detection: A DFT study. *Sens. Actuators B* **2015**, *212*, 55–62. [CrossRef]

7. Mu, H.; Wang, K.; Zhang, S.; Shi, K.; Sun, S.; Li, Z.; Zhou, J.; Xie, H. Fabrication and characterization of amino group functionalized multiwall carbon nanotubes (MWCNT) formaldehyde gas sensors. *IEEE Sens. J.* **2014**, *14*, 2362–2368. [CrossRef]

8. Xie, H.; Sheng, C.; Chen, X.; Wang, X.; Li, Z.; Zhou, J. Multi-wall carbon nanotube gas sensors modified with amino-group to detect low concentration of formaldehyde. *Sens. Actuators B* **2012**, *168*, 34–38. [CrossRef]

9. Antwi-Boampong, S.; Peng, J.S.; Carlan, J.; Belbruno, J.J. A molecularly imprinted fluoral-p/polyaniline double layer sensor system for selective sensing of formaldehyde. *IEEE Sens. J.* **2014**, *14*, 1490–1498. [CrossRef]

10. Ruppel, C.C.W. Acoustic wave filter technology-A review. *IEEE Trans. Ultrason. Ferroelectr. Freq. Control* **2017**, *64*, 1390–1400. [CrossRef] [PubMed]

11. Hashimoto, K.Y. Radio frequency bulk acoustic wave filters: Current status and future prospective. *IEEE Trans. Electron. Inf. Syst.* **2013**, *133*, 502–508. [CrossRef]

12. Zhu, Y.; Wang, N.; Sun, C.; Merugu, S.; Singh, N.; Gu, Y. A High Coupling Coefficient 2.3-GHz AlN Resonator for High Band LTE Filtering Application. *IEEE Electron Device Lett.* **2016**, *37*, 1344–1346. [CrossRef]

13. Wang, J.; Chen, D.; Gan, Y.; Sun, X.; Jin, Y. High sensitive self-assembled monolayer modified solid mounted resonator for organophosphate vapor detection. *Appl. Surf. Sci.* **2011**, *257*, 4365–4369. [CrossRef]

14. Zhao, X.; Ashley, G.M.; Garcia-Gancedo, L.; Jin, H.; Luo, J.; Flewitt, A.J.; Lu, J.R. Protein functionalized ZnO thin film bulk acoustic resonator as an odorant biosensor. *Sens. Actuators B* **2012**, *163*, 242–246. [CrossRef]

15. Chen, D.; Wang, J.J.; Li, D.H.; Xu, Y. Hydrogen sensor based on Pd-functionalized film bulk acoustic resonator. *Sens. Actuators B* **2011**, *159*, 234–237. [CrossRef]

16. Chang, Y.; Tang, N.; Qu, H.; Liu, J.; Zhang, D.; Zhang, H.; Pang, W.; Duan, X. Detection of volatile organic compounds by self-assembled monolayer coated sSensor array with concentration-independent fingerprints. *Sci. Rep.* **2016**, *6*. [CrossRef]

17. Zhao, X.; Pan, F.; Ashley, G.M.; Garcia-Gancedo, L.; Luo, J.; Flewitt, A.J.; Milne, W.I.; Lu, J.R. Label-free detection of human prostate-specific antigen (HPSA) using film bulk acoustic resonators (FBARs). *Sens. Actuators B* **2014**, *190*, 946–953. [CrossRef]

18. Chen, D.; Wang, J.; Wang, P.; Guo, Q.; Zhang, Z.; Ma, J. Real-time monitoring of human blood clotting using a lateral excited film bulk acoustic resonator. *J. Micromech. Microeng.* **2017**, *27*, 045013. [CrossRef]

19. Voiculescu, I.; Nordin, A.N. Acoustic wave based MEMS devices for biosensing applications. *Biosens. Bioelectron.* **2012**, *33*, 1–9. [CrossRef] [PubMed]

20. Chen, D.; Song, S.; Ma, J.; Zhang, Z.; Wang, P.; Liu, W.; Guo, Q. Micro-electromechanical film bulk acoustic sensor for plasma and whole blood coagulation monitoring. *Biosens. Bioelectron.* **2017**, *91*, 465–471. [CrossRef] [PubMed]

21. Pang, W.; Zhao, H.; Kim, E.S.; Zhang, H.; Yuc, H.; Hu, X. Piezoelectric microelectromechanical resonant sensors for chemical and biological detection. *Lab Chip* **2012**, *12*, 29–44. [CrossRef] [PubMed]

22. Chen, D.; Wang, J.; Xu, Y.; Li, D. A pure shear mode ZnO film resonator for the detection of organophosphorous pesticides. *Sens. Actuators B* **2012**, *171–172*, 1081–1086. [CrossRef]

23. Chen, D.; Zhang, Z.; Ma, J.; Wang, W. ZnO Film Bulk Acoustic Resonator for the Kinetics Study of Human Blood Coagulation. *Sensors* **2017**, *17*, 1015. [CrossRef] [PubMed]

24. Chen, D.; Wang, J.; Li, D.; Liu, Y.; Song, H.; Liu, Q. A poly(vinylidene fluoride)-coated ZnO film bulk acoustic resonator for nerve gas detection. *J. Micromech. Microeng.* **2011**, *21*, 085017–085023. [CrossRef]

25. Wang, J.-J.; Liu, W.-H.; Chen, D.; Xu, Y.; Zhang, L.-Y. A micro-machined thin film electro-acoustic biosensor for detection of pesticide residuals. *J. Zhejiang Univ.-Sci. C-Comput. Electron.* **2014**, *15*, 383–389. [CrossRef]

26. Zhang, M.; Huang, J.; Cui, W.; Pang, W.; Zhang, H.; Zhang, D.; Duan, X. Kinetic studies of microfabricated biosensors using local adsorption strategy. *Biosens. Bioelectron.* **2015**, *74*, 8–15. [CrossRef] [PubMed]

27. Wang, J.; Chen, D.; Xu, Y.; Liu, W. Label-free immunosensor based on micromachined bulk acoustic resonator for the detection of trace pesticide residues. *Sens. Actuators B* **2014**, *190*, 378–383. [CrossRef]

28. Guo, P.; Xiong, J.; Zheng, D.; Zhang, W.; Liu, L.; Wang, S.; Gu, H. A biosensor based on a film bulk acoustic resonator and biotin-avidin system for the detection of the epithelial tumor marker mucin 1. *RSC Adv.* **2015**, *5*, 66355–66359. [CrossRef]

29. Campos, J.; Jiménez, C.; Trigo, C.; Ibarra, P.; Rana, D.; Thiruganesh, R.; Ramalingam, M.; Haidar, Z.S. Quartz crystal microbalance with dissipation monitoring: A powerful tool for bionanoscience and drug discovery. *J. Bionanosci.* **2015**, *9*, 249–260. [CrossRef]

30. Wu, Z.; Chen, X.; Zhu, S.; Zhou, Z.; Yao, Y.; Quan, W.; Liu, B. Enhanced sensitivity of ammonia sensor using graphene/polyaniline nanocomposite. *Sens. Actuators B* **2013**, *178*, 485–493. [CrossRef]

31. Khassaf, H.; Khakpash, N.; Sun, F.; Sbrockey, N.M.; Tompa, G.S.; Kalkur, T.S.; Alpay, S.P. Strain engineered barium strontium titanate for tunable thin film resonators. *Appl. Phys. Lett.* **2014**, *104*, 202902. [CrossRef]

32. Chen, D.; Wang, J.; Xu, Y.; Li, D.; Zhang, L.; Li, Z. Highly sensitive detection of organophosphorus pesticides by acetylcholinesterase-coated thin film bulk acoustic resonator mass-loading sensor. *Biosens. Bioelectron.* **2013**, *41*, 163–167. [CrossRef] [PubMed]

33. Chen, D.; Wang, J.; Xu, Y. Highly sensitive lateral field excited piezoelectric film acoustic enzyme biosensor. *IEEE Sens. J.* **2013**, *13*, 2217–2222. [CrossRef]

34. Lu, Y.; Chang, Y.; Tang, N.; Qu, H.; Liu, J.; Pang, W.; Zhang, H.; Zhang, D.; Duan, X. Detection of volatile organic compounds using microfabricated resonator array functionalized with supramolecular monolayers. *ACS Appl. Mater. Interfaces* **2015**, *7*, 17893–17903. [CrossRef] [PubMed]

35. Chen, D.; Wang, J.; Li, D.; Xu, Y.; Li, Z. Solidly mounted resonators operated in thickness shear mode based on c-axis oriented AlN films. *Sens. Actuators A* **2011**, *165*, 379–384. [CrossRef]

36. Jingjing, W.; Da, Z.; Ke, W.; Weiwei, H. The detection of formaldehyde using microelectromechanical acoustic resonator with multiwalled carbon nanotubes-polyethyleneimine composite coating. *J. Micromech. Microeng.* **2018**, *28*, 015003.

37. Zhang, C.; Wang, X.; Lin, J.; Ding, B.; Yu, J.; Pan, N. Nanoporous polystyrene fibers functionalized by polyethyleneimine for enhanced formaldehyde sensing. *Sens. Actuators B* **2011**, *152*, 316–323. [CrossRef]

38. Wang, X.; Ding, B.; Sun, M.; Yu, J.; Sun, G. Nanofibrous polyethyleneimine membranes as sensitive coatings for quartz crystal microbalance-based formaldehyde sensors. *Sens. Actuators B* **2010**, *144*, 11–17. [CrossRef]

39. Wang, X.; Cui, F.; Lin, J.; Ding, B.; Yu, J.; Al-Deyab, S.S. Functionalized nanoporous TiO₂ fibers on quartz crystal microbalance platform for formaldehyde sensor. *Sens. Actuators B* **2012**, *171–172*, 658–665. [CrossRef]

40. Tai, H.; Bao, X.; He, Y.; Du, X.; Xie, G.; Jiang, Y. Enhanced formaldehyde-sensing performances of mixed polyethyleneimine-multiwalled carbon nanotubes composite films on quartz crystal microbalance. *IEEE Sens. J.* **2015**, *15*, 6904–6911. [CrossRef]

41. Li, L.; Xiang, Y.; Shang, Y. *Basic Origanic Chemistry*; Tsinghua University Press: Beijing, China, 2008; pp. 350–351. ISBN 9787302186571.

42. Zhang, H.; Marma, M.S.; Kim, E.S.; McKenna, C.E. Thompson, M.E. A film bulk acoustic resonator in liquid environments. *J. Micromech. Microeng.* **2005**, *15*, 1911–1916. [CrossRef]

43. Chen, D.; Song, S.; Zhang, D.; Wang, P.; Liu, W. Sensing characteristics of pure-shear film bulk acoustic resonator in viscous liquids. *Mod. Phys. Lett. B* **2017**, *31*, 1750086. [CrossRef]

44. Wang, P.; Su, J.; Dai, W.; Cernigliaro, G.; Sun, H. Ultrasensitive quartz crystal microbalance enabled by micropillar structure. *Appl. Phys. Lett.* **2014**, *104*, 043504. [CrossRef]

45. Wang, P.; Su, J.; Su, C.F.; Dai, W.; Cernigliaro, G.; Sun, H. An ultrasensitive quartz crystal microbalance-micropillars based sensor for humidity detection. *J. Appl. Phys.* **2014**, *115*, 224501. [CrossRef]

46. Chung, P.R.; Tzeng, C.T.; Ke, M.T.; Lee, C.Y. Formaldehyde gas sensors: A review. *Sensors* **2013**, *13*, 4468–4484. [CrossRef] [PubMed]

47. Pramod, N.G.; Pandey, S.N.; Sahay, P.P. Sn-doped In₂O₃ nanocrystalline thin films deposited by spray pyrolysis: Microstructural, optical, electrical, and formaldehyde-sensing characteristics. *J. Therm. Spray Technol.* **2013**, *22*, 1035–1043. [CrossRef]

48. Mansfeld, G.D. Theory of high overtone bulk acoustic wave resonator as a gas sensor. In Proceedings of the 13th International Conference on Microwaves, Radar and Wireless Communications, MIKON 2000, Wroclaw, Poland, 22–24 May 2000; pp. 469–472.

49. Cranston, E.D.; Eita, M.; Johansson, E.; Netrval, J.; Salajková, M.; Arwin, H.; Wagberg, L. Determination of Young's modulus for nanofibrillated cellulose multilayer thin films using buckling mechanics. *Biomacromolecules* **2011**, *12*, 961–969. [CrossRef] [PubMed]

50. Penza, M.; Aversa, P.; Cassano, G.; Suriano, D.; Wlodarski, W.; Benetti, M.; Cannatà, D.; Pietrantonio, F.D.; Verona, E. Thin-film bulk-acoustic-resonator gas sensor functionalized with a nanocomposite langmuir-blodgett layer of carbon nanotubes. *IEEE Trans. Electron Devices* **2008**, *55*, 1237–1243. [CrossRef]

51. Dimitrova, Y. Solvent effects on vibrational spectra of hydrogen-bonded complexes of formaldehyde and water: An ab initio study. *J. Mol. Struct. THEOCHEM* **1997**, *391*, 251–257. [CrossRef]

micromachines

MDPI

Article

Monostable Dynamic Analysis of Microbeam-Based Resonators via an Improved One Degree of Freedom Model

Lei Li [1,2,*], Qichang Zhang [2], Wei Wang [2] and Jianxin Han [3]

[1] School of Transportation and Vehicle Engineering, Shandong University of Technology, Zibo 255049, China
[2] Department of Mechanics, School of Mechanical Engineering, Tianjin University, Tianjin 300350, China; qzhang@tju.edu.cn (Q.Z.); wangweifrancis@tju.edu.cn (W.W.)
[3] Tianjin Key Laboratory of High Speed Cutting and Precision Machining, Tianjin University of Technology and Education, Tianjin 300222, China; hanjianxin@tju.edu.cn
* Correspondence: lileizi888@tju.edu.cn

Received: 17 December 2017; Accepted: 19 February 2018; Published: 22 February 2018

Abstract: Monostable vibration can eliminate dynamic bifurcation and improve system stability, which is required in many microelectromechanical systems (MEMS) applications, such as microbeam-based and comb-driven resonators. This article aims to theoretically investigate the monostable vibration in size-effected MEMS via a low dimensional model. An improved single degree of freedom model to describe electrically actuated microbeam-based resonators is obtained by using modified couple stress theory and Nonlinear Galerkin method. Static displacement, pull-in voltage, resonant frequency and especially the monostable dynamic behaviors of the resonators are investigated in detail. Through perturbation analysis, an approximate average equation is derived by the application of the method of Multiple Scales. Theoretical expressions about parameter space and maximum amplitude of monostable vibration are then deduced. Results show that this improved model can describe the static behavior more accurately than that of single degree of freedom model via traditional Galerkin Method. This desired monostable large amplitude vibration is significantly affected by the ratio of the gap width to mircobeam thickness. The optimization design results show that reasonable decrease of this ratio can be beneficial to monostable vibration. All these analytical results are verified by numerical results via Differential Quadrature method, which show excellent agreement with each other. This analysis has the potential of improving dynamic performance in MEMS.

Keywords: MEMS; monostable vibration; Nonlinear Galerkin method; optimization

1. Introduction

Microbeam-based structures are widely applied in MEMS, such as microactuator/sensor [1–3], energy harvester [4], microresonator [5–7], gyroscope [8], microgripper [9,10] and so on. Their light weight, small size, low-energy consumption and durability make them even more attractive. MEMS pressure sensors and accelerometers are widely used in the automotive industry. Some microsensors and actuators are also adopted for various biomedical applications. In general, the operation of these electrostatically actuated resonant devices is based on linear resonance [11]. However, these dynamic systems are nonlinear and the output energy is very small in the case of linear resonance, which is undesirable in MEMS. Therefore, monostable large amplitude vibration is required in MEMS sensing application. It can eliminate dynamic bifurcation phenomenon and improve system stability. Zhang et al. [12] studied the dynamic behavior of comb-driven resonators with inclination of the fingers and edge effect on the capacitance and obtain parameters for linear resonance operation.

The inclination of the fingers can be beneficial to restrain electrostatic nonlinearity and make the system realize linear vibration. Han et al. [13] studied dynamic behaviors of a doubly clamped microresonator and presented some design considerations to realize large amplitude vibration. Masri et al. [14] studied the stability of MEMS resonators when undergoing large amplitude motion with a delayed feedback velocity controller.

With the reduction of the scale of the MEMS, classic continuum mechanics theories are unable to describe the size effects [15], thus higher-order continuum theories or non-classic theories are inevitably needed to describe scale effect and corresponding relations in continuum mechanics [16]. The couple stress theory is considered as one of the higher-order continuum theories [17]. It involves parameters for expressing the effect of material length scale which possess the capability to describe the size effect of microstructures [18]. This theory includes two extra material length scale parameters besides the two classic material constants for elastic isotropic materials. Anthoine [19] and Yang et al. [20] presented the modified couple stress theory in which the two material length scale parameters are decreased to only one parameter. This feature facilitates the use of the modified couple stress theory for the study of micro- and nano-scale structures.

Static/dynamic behaviors of electrically actuated microbeams have been studied a lot by using different models and approaches. These investigations can be divided into two groups. The first group focuses on lumped-mass model and establishes single degree of freedom equation. Then, perturbation method is introduced to study dynamic behaviors of microbeam, which provides theoretical guidance for engineering [21–24]. However, lumped-mass model cannot accurately describe dynamic characteristics of mircobeam as the increase of amplitude. The other group focuses on continuum model and establishes partial differential equations which are then solved with a variety of numerical methods, such as Galerkin discretization and Differential Quadrature method. These results can describe dynamic characteristics of mircobeam more accurately [11,25–27]. Although the continuum model has high accuracy, it is not conducive to theoretical analysis in depth. Based on the continuum model, Younis et al. [28–34] studied static pull-in behavior and dynamic pull-in behavior of electrically actuated microbeams by using the Galerkin method, the Differential Quadrature method and the Shooting method. Results showed that the single degree of freedom model cannot accurately describe static displacement, pull-in voltage, resonant frequency or vibration amplitude. Specially, when amplitude exceeds half of the gap, the error between single degree of freedom model and continuum model increases significantly. Besides, Nayfeh et al. [23] analyzed the vibration behaviors of a Euler-Bernoulli beam with direct application of the method of multiple scales to the governing partial-differential equation. However, high-order vibration items were not considered in their study. In fact, high-order vibration items have important influence on dynamic behavior [25]. Fortunately, the Nonlinear Galerkin method can solve it with introducing higher modes into the single degree of freedom model [35].

The Nonlinear Galerkin method and the well-known Galerkin method can be used to obtain the low dimensional manifold by some projection onto a sub manifold [35,36]. However, the well-known Galerkin method restrict the sub-manifold at being a flat sub-manifold; the Nonlinear Galerkin method tries to improve on this by not restricting the sub-manifold to an affine sub-space. Kang et al. [37] studied dynamic behaviors of low-dimensional modeling of the fluid dynamic system with the Nonlinear Galerkin method. Considering the effect of higher modes, the Nonlinear Galerkin method can give an accurate description for the dynamic behaviors of the system. With the Nonlinear Galerkin method, partial differential equations can be discretized into a finite-degree-of-freedom system consisting of ordinary-differential equations and then, a low-dimensional model containing higher modes information can be generated.

It can be concluded from the above analysis that monostable vibration can eliminate dynamic bifurcation, improve system stability and increase the output of energy, which is desired in many MEMS applications. However, to the best of our knowledge, there are fewer effective methods to study monostable large amplitude vibration behaviors via the low dimensional model. This paper aims to obtain an improved single degree of freedom model by using modified couple stress theory and

Micromachines **2018**, *9*, 89

Nonlinear Galerkin method and deduce theoretical expressions about parameter space and maximum amplitude of monostable vibration.

The rest of this paper is organized as follows. In Section 2, a novel single degree of freedom model to describe electrically actuated microbeam-based resonators is obtained by using modified couple stress theory and Nonlinear Galerkin method. In Section 3, the method of Multiple Scales is used to derive an approximate average equation. In Section 4, static and dynamic properties of these devices are then investigated in detail. Parameter space and maximum amplitude of the monostable vibration are theoretically derived and numerically verified. Concluding remarks are given in the last section.

2. Mathematical Model

2.1. Governing Equation

Here, we consider a clamped-clamped microbeam-based resonator, as shown in Figure 1. The actuation of the microbeam is realized by means of a bias voltage and an AC voltage component. The microbeam and the electrode are made from silicon material. Based on the modified couple stress theory, the equation of motion that governs the transverse deflection $\hat{w}(x,t)$ is written as [17].

$$\rho A\ddot{\hat{w}} + (EI + \mu Al^2)\hat{w}^{iv} + c\dot{\hat{w}} = \left(\hat{N} + \frac{EA}{2L}\int_0^L \hat{w}'^2 d\hat{x}\right)\hat{w}'' + \frac{\varepsilon_0 b[V_{dc} + V_{ac}\cos(\hat{\Omega}t)]^2}{2(d - \hat{w})^2} \tag{1}$$

with the following boundary conditions

$$\hat{w}(0,\hat{t}) = \hat{w}'(0,\hat{t}) = \hat{w}(L,\hat{t}) = \hat{w}'(L,\hat{t}) = 0 \tag{2}$$

where $\dot{\hat{w}} = \frac{\partial \hat{w}}{\partial \hat{t}}$ and $\hat{w}' = \frac{\partial \hat{w}}{\partial \hat{x}}$.

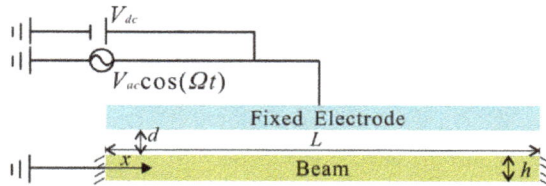

Figure 1. Schematic of an electrically actuated microbeam.

The first term on the right hand of Equation (1) represents the axial force and mid-plane stretching effects. Here, $\hat{x}$ is the position along the microbeam length; A and I are the area and moment of inertia of the cross section; d is the gap width; ε_0 is the dielectric constant of the gap medium. The parameter N corresponds to a tensile or compressive axial load, depending on whether it is positive or negative. l is introduced into Equation (1) as the material length scale parameter that has the capability to physically model properties of the couple stress effect. The last term in Equation (1) represents the parallel-plate electric actuation which is composed of DC and AC components. Here, DC voltage can cause a static deflection in the microbeam. There is a limit for the applied DC voltage called the static pull-in voltage [38]. AC voltage, which is small compared to DC voltage, causes the dynamic response of microbeam. Part of system parameters are defined as stated in Table 1.

Table 1. Part of design parameters for a microbeam-based resonator.

Parameter	Value	Units
Mass density ρ	2300	kg/m^3
Young's modulus E	169	Gpa
Beam length L	365	μm
Beam width b	10	μm
Beam thickness h	1	μm
Axial load $\hat{N}$	variable	N
Viscous damping c	3.42×10^{-5}	Ns/m^2

In the relations above, μ is Lame's constant that is defined by Young's modulus E and Poisson's ratio v as

$$\mu = \frac{E}{2(1+v)} \tag{3}$$

For convenience, the following non-dimensional variables are introduced

$$w = \frac{\hat{w}}{d}, \; x = \frac{\hat{x}}{L}, \; t = \hat{t}\sqrt{\frac{EI}{\rho AL^4}} \tag{4}$$

Substituting the non-dimensional variables into Equations (1) and (2), yields the following non-dimensional equation of motion of the micro-resonator

$$\ddot{w} + (1+\eta)w^{iv} + c_n\dot{w} - (N + \alpha_1\int_0^1 w'^2 dx)w'' = \alpha_2\frac{(V_{dc} + V_{ac}\cos\Omega t)^2}{(1-w)^2} \tag{5}$$

with boundary conditions

$$w(0,t) = w'(0,t) = w(1,t) = w'(1,t) = 0 \tag{6}$$

The parameters appearing in Equation (5) are

$$\alpha_1 = 6 \times (\frac{d}{h})^2, \alpha_2 = \frac{6\varepsilon_0 L^4}{Ed^3h^3}, \eta = \frac{\mu A l^2}{EI}, N = \frac{\hat{N}L^2}{EI} \tag{7}$$

where α_1 represents ratio coefficient of the gap width to the mircobeam thickness, α_2 represents electrostatic force coefficient and η represents scale effect.

2.2. The Nonlinear Galerkin Method

Compared with the Linear Galerkin method, the Nonlinear Galerkin method can describe the dynamic behaviors of the system more accurately. In this section, we introduce the Nonlinear Galerkin method to deal with Equation (5) and obtain an improved one degree of freedom model [36].

Firstly, considering the Galerkin procedure, we discretize Equation (5) into a finite-degree-of-freedom system consisting of ordinary differential equations. This technique and its application to nonlinear systems were discussed by Abdel-Rahman et al. [39].

The solution of Equation (5) can be expressed as

$$w(x,t) = \sum_{i=1}^{\infty} u_i(t)\phi_i(x) \tag{8}$$

where ϕ_i is the i-th linear undamped mode shape of the straight microbeam, normalized such that $\int_0^1 \phi_i \phi_j dx = \delta_{ij}$ and governed by

$$(1+\eta)\phi_i^{iv} = N\phi_i'' + \omega_i^2 \phi_i \tag{9}$$

with

$$\phi_i(0) = \phi_i(1) = \phi_i'(0) = \phi_i'(1) = 0 \tag{10}$$

where ω_i is the i-th natural frequency of the microbeam. Equations (9) and (10) represent a boundary-value problem that can be solved by using a combination of Shooting method and a bisection procedure for each pair of mode shape and natural frequency [39].

Then, Equation (5) is multiplied by $(1-w)^2$ [25], so that the electric-force term is represented exactly. Substituting Equation (8) into the resulting equation, multiplying by ϕ_i and integrating the outcome from $x = 0$–1, yield

$$
\begin{aligned}
&\ddot{u}_n + \omega_n^2 u_n + c_n \dot{u}_n - 2\sum_{i,j=1}^{M} u_i \ddot{u}_j \int_0^1 \phi_i \phi_j \phi_n dx + \sum_{i,j,k=1}^{M} u_i u_j \ddot{u}_k \int_0^1 \phi_i \phi_j \phi_k \phi_n dx \\
&= 2\sum_{i,j=1}^{M} \omega_i^2 u_i u_j \int_0^1 \phi_i \phi_j \phi_n dx - \sum_{i,j,k=1}^{M} \omega_i^2 u_i u_j u_k \int_0^1 \phi_i \phi_j \phi_k \phi_n dx + 2c_n \sum_{i,j=1}^{M} u_i \dot{u}_j \int_0^1 \phi_i \phi_j \phi_n dx \\
&- c_n \sum_{i,j,k=1}^{M} u_i u_j \dot{u}_k \int_0^1 \phi_i \phi_j \phi_k \phi_n dx + \alpha_1 \sum_{i,j,k=1}^{M} u_i u_j u_k \Gamma(\phi_i,\phi_j) \int_0^1 \phi_k'' \phi_n dx \\
&- 2\alpha_1 \sum_{i,j,k,l=1}^{M} u_i u_j u_k u_l \Gamma(\phi_i,\phi_j) \int_0^1 \phi_k \phi_l'' \phi_n dx + \alpha_1 \sum_{i,j,k,l,m=1}^{M} u_i u_j u_k u_l u_m \Gamma(\phi_i,\phi_j) \int_0^1 \phi_k \phi_l \phi_m'' \phi_n dx \\
&+ \alpha_2 (V_{dc} + V_{ac} \cos \Omega t)^2 \int_0^1 \phi_n dx
\end{aligned}
\tag{11}
$$

where $\Gamma(\phi_i, \phi_j) = \int_0^1 \phi_i \phi_j dx$. Due to $V_{dc} >> V_{ac}$ [34], $(V_{dc} + V_{ac}\cos\Omega t)^2 \approx V_{dc}^2 + 2V_{dc}V_{ac}\cos\Omega t$ is obtained.

Equation (11) represents a discretized system consisting of ordinary-differential equations, which contain all nonlinearities up to fifth order. The above is the Linear Galerkin method for dimensionality reduction. Most researchers utilized the derived reduced-order models to simulate the static behavior and dynamic response of microbeam-based MEMS devices. However, the way is not conducive to theoretical analysis. Here, the Nonlinear Galerkin method is introduced to deal with the above equations.

Starting with an abstract setting, a nonlinear dynamical system is separated into the linear and the higher order nonlinear part

$$\dot{x} + g(x,t) = \dot{x} + Ax + h(x,t) = 0, x \in R^{2 \times M} \tag{12}$$

with Ax as linear part and $h(x,t)$ as nonlinear part of the system $g(x,t)$. Here we take $2M$ spatial dimension corresponding with Equation (11).

Here, we assume the solution x as

$$x = Y_m \xi + Z_m \eta \tag{13}$$

where the columns of the matrix $Y_m = [y_1, \cdots, y_m]$ span the m-dimensional sub-space span $\{y_1, \cdots, y_m\}$ and the columns of the matrix $Z_m = [y_{m+1}, \cdots, y_{2 \times M}]$ span complementary sub-space span $\{y_{m+1}, \cdots, y_{2 \times M}\}$.

Substituting Equation (13) into Equation (12), multiplying by $\widetilde{Y}_m^T$ and $\widetilde{Z}_m^T$ from the left, yield

$$
\begin{aligned}
\dot{\xi} + \widetilde{Y}^T A Y \xi + \widetilde{Y}^T h(Y\xi + Z\eta, t) &= 0 \\
\dot{\eta} + \widetilde{Z}^T A Z \eta + \widetilde{Z}^T h(Y\xi + Z\eta, t) &= 0
\end{aligned}
\tag{14}
$$

where $\widetilde{Y}_m^T Y_m = \mathbf{I}$, $\widetilde{Z}_m^T Z_m = \mathbf{I}$ and $\mathbf{I}$ is unit matrix.

Combination with the dynamic response of microbeam, ξ represents low-frequency part and η represents high-frequency part. To describe dynamic response of system accurately with the low-frequency part, the following relation is introduced

$$\eta = \phi(\xi) \tag{15}$$

Substituting Equation (15) into Equation (14), we can obtain

$$\dot{\xi} + \widetilde{Y}^T A Y \xi + \widetilde{Y}^T h(Y\xi + Z\phi(\xi), t) = 0 \tag{16}$$

Both low-frequency part and high-frequency part are introduced in Equation (16). It is considered that the high-frequency vibration has little impact on system dynamics under primary resonance condition. So, we set $\dot{\eta} = 0$. Then, we can obtain the reduced order model that contains the information of high order modes.

To obtain dynamic equation of single degree of freedom, we set subspace dimension equal to 2.

In order to simplify the calculation, we set M equal to 3 and only keep linear part of the high dimensional space variables. The relationship between the high-frequency part and low-frequency part is obtained, as shown below

$$
\begin{aligned}
u_3 = \frac{1}{\omega_3^2} [& 2\int_0^1 \phi_1^2 \phi_3 dx (u_1 \ddot{u}_1 + c_n u_1 \dot{u}_1 + \omega_1^2 u_1^2) - u_1^2 \ddot{u}_1 \int_0^1 \phi_1^3 \phi_3 dx - \omega_1^2 u_1^3 \int_0^1 \phi_1^3 \phi_3 dx - c_n u_1^2 \dot{u}_1 \int_0^1 \phi_1^3 \phi_3 dx \\
& + \alpha_1 u_1^3 \int_0^1 \phi_1'^2 dx \int_0^1 \phi_1'' \phi_3 dx - 2\alpha_1 u_1^4 \int_0^1 \phi_1'^2 dx \int_0^1 \phi_1'' \phi_3 \phi_1 dx + \alpha_2 V_{dc} \int_0^1 \phi_3 dx]
\end{aligned} \tag{17}
$$

Substituting Equation (17) into Equation (11) and keeping all nonlinearities up to fifth order, yield the following novel single degree of freedom equation

$$
\begin{aligned}
(\ddot{u}_1 + \omega_1^2 u_1 + c_n \dot{u}_1)(n_0 + n_1 u_1 + n_2 u_1^2 + n_3 u_1^3 + n_4 u_1^4) \\
= m_1 u_1^3 + m_2 (u_1^2 \ddot{u}_1 + c_n u_1^2 \dot{u}_1) + m_3 u_1 + m_4 + 2 m_5 V_{dc} V_{ac} \cos \Omega t \\
+ m_6 u_1^2 + m_7 u_1^4 + m_8 (u_1^3 \ddot{u}_1 + c_n u_1^3 \dot{u}_1) + m_9 u_1^5
\end{aligned} \tag{18}
$$

where coefficients are expressed as Equations (A1)–(A15) in Appendix A.

The above is the Nonlinear Galerkin method for dimensionality reduction. The previous single degree of freedom model obtained by the Linear Galerkin method only contains the first mode. In this paper, the Nonlinear Galerkin method is introduced to obtain the improved single degree of freedom model that contains the first mode and the third mode, which improves the accuracy of the model.

This paper aims to study monostable large amplitude vibration. With the increase of amplitude, the advantage of the improved model becomes more and more obvious.

3. Perturbation Analysis

Static analysis is significant in the MEMS. Through it, equilibrium position, pull-in voltage and pull-in location of the system can be obtained. Each of the DC voltage corresponds to a static displacement and the perturbation method is used to obtain an approximate solution near the equilibrium position. We assume $u_1 = u_{1d} + u_{1s}$, where u_{1d} and u_{1s} represent dynamic behavior and static behavior, respectively. Then, we can obtain the static equation by making the u_1 independent of time

$$\omega_1^2 u_{1s}(n_0 + n_1 u_{1s} + n_2 u_{1s}^2 + n_3 u_{1s}^3 + n_4 u_{1s}^4) = m_1 u_{1s}^3 + m_3 u_{1s} + m_4 + m_6 u_{1s}^2 + m_7 u_{1s}^4 + m_9 u_{1s}^5 \tag{19}$$

and the dynamic equation by ignoring high order damping terms

$$
\begin{aligned}
\ddot{u}_{1d} + \omega_{1d}^2 u_{1d} + c_n \dot{u}_{1d} = & \; q_1 u_{1d} \ddot{u}_{1d} + q_2 u_{1d}^2 \ddot{u}_{1d} + q_3 u_{1d}^3 \ddot{u}_{1d} + q_4 u_{1d}^4 \ddot{u}_{1d} + q_5 u_{1d}^2 \\
& + q_6 u_{1d}^3 + q_7 u_{1d}^4 + q_8 u_{1d}^5 + q_9 V_{ac} \cos \Omega t
\end{aligned} \tag{20}
$$

where coefficients are expressed as Equations (A16)–(A25) in Appendix A.

To indicate the significance of each term in the equation of motion, ε is introduced as a small non-dimensional bookkeeping parameter. Considering the electrostatic force term $q_9 = O(\varepsilon^5)$, scaling the dissipative terms, we obtain

$$
\begin{aligned}
\ddot{u}_{1d} + \omega_{1d}^2 u_{1d} + \varepsilon^4 c_n \dot{u}_{1d} &= q_1 u_{1d} \ddot{u}_{1d} + q_2 u_{1d}^2 \ddot{u}_{1d} + q_3 u_{1d}^3 \ddot{u}_{1d} + q_4 u_{1d}^4 \ddot{u}_{1d} + q_5 u_{1d}^2 \\
&+ q_6 u_{1d}^3 + q_7 u_{1d}^4 + q_8 u_{1d}^5 + \varepsilon^5 q_9 V_{ac} \cos \Omega t
\end{aligned}
\tag{21}
$$

To express the relationship between the excitation frequency and the natural frequency, we introduce a detuning parameter σ defined by $\Omega = \omega_{1d} + \varepsilon^4 \sigma$. Here, σ is the tuning parameter. Then, to determine a fifth-order uniform expansion of the solution of Equation (21) by using the method of Multiple Scales, we introduce three time scales $T_0 = t$, $T_2 = \varepsilon^2 t$ and $T_4 = \varepsilon^4 t$ [23] and expand the time-dependent variable u_{1d} in powers of ε as

$$
u_{1d} = \varepsilon v_1(T_0, T_2, T_4) + \varepsilon^2 v_2(T_0, T_2, T_4) + \varepsilon^3 v_3(T_0, T_2, T_4) + \varepsilon^4 v_4(T_0, T_2, T_4) + \varepsilon^5 v_5(T_0, T_2, T_4)
\tag{22}
$$

Substituting Equation (22) into Equation (21) and equating coefficients of like powers of ε, yields

order ε:

$$
\frac{\partial^2 v_1}{\partial T_0^2} + \omega_{1d}^2 v_1 = 0
\tag{23}
$$

order ε^2:

$$
\frac{\partial^2 v_2}{\partial T_0^2} + \omega_{1d}^2 v_2 = q_1 v_1 \frac{\partial^2 v_1}{\partial T_0^2} + q_5 v_1^2
\tag{24}
$$

order ε^3:

$$
\frac{\partial^2 v_3}{\partial T_0^2} + \omega_{1d}^2 v_3 = -2 \frac{\partial^2 v_1}{\partial T_0 \partial T_2} + q_1 v_2 \frac{\partial^2 v_1}{\partial T_0^2} + q_1 v_1 \frac{\partial^2 v_2}{\partial T_0^2} + 2 q_5 v_2 v_1 + q_2 v_1^2 \frac{\partial^2 v_1}{\partial T_0^2} + q_6 v_1^3
\tag{25}
$$

order ε^4:

$$
\begin{aligned}
\frac{\partial^2 v_4}{\partial T_0^2} + \omega_{1d}^2 v_4 &= -2 D_0 D_2 v_2 + q_1 \left(v_1 \frac{\partial^2 v_3}{\partial T_0^2} + v_3 \frac{\partial^2 v_1}{\partial T_0^2} + v_2 \frac{\partial^2 v_2}{\partial T_0^2} + 2 v_1 \frac{\partial^2 v_1}{\partial T_0 \partial T_2} \right) \\
&+ q_2 \left(2 v_2 v_1 \frac{\partial^2 v_1}{\partial T_0^2} + v_1^2 \frac{\partial^2 v_2}{\partial T_0^2} \right) + q_3 v_1^3 \frac{\partial^2 v_1}{\partial T_0^2} + q_5 (2 v_1 v_3 + v_2^2) + 3 q_6 v_1^2 v_2 + q_7 v_1^4
\end{aligned}
\tag{26}
$$

order ε^5:

$$
\begin{aligned}
\frac{\partial^2 v_5}{\partial T_0^2} + \omega_{1d}^2 v_5 &= -2 D_0 D_4 v_1 - 2 D_0 D_2 v_3 - D_2 D_2 v_1 - c_n \frac{\partial v_1}{\partial T_0} + q_1 \left(v_1 \frac{\partial^2 v_4}{\partial T_0^2} + v_4 \frac{\partial^2 v_1}{\partial T_0^2} \right. \\
&\left. + v_3 \frac{\partial^2 v_2}{\partial T_0^2} + v_2 \frac{\partial^2 v_3}{\partial T_0^2} + 2 v_1 \frac{\partial^2 v_3}{\partial T_0 \partial T_2} + 2 v_2 \frac{\partial^2 v_1}{\partial T_0 \partial T_2} \right) + 4 q_7 v_1^3 v_2 + q_8 v_1^5 + q_2 \left(2 v_3 v_1 \frac{\partial^2 v_1}{\partial T_0^2} \right. \\
&\left. + v_1^2 \frac{\partial^2 v_3}{\partial T_0^2} + 2 v_2 v_1 \frac{\partial^2 v_2}{\partial T_0^2} + v_2^2 \frac{\partial^2 v_1}{\partial T_0^2} \right) + q_3 \left(v_1^3 \frac{\partial^2 v_2}{\partial T_0^2} + 3 v_2 v_1^2 \frac{\partial^2 v_1}{\partial T_0^2} \right) + q_4 v_1^4 \frac{\partial^2 v_1}{\partial T_0^2} \\
&+ q_5 (2 v_1 v_4 + 2 v_2 v_3) + 3 q_6 (v_2^2 v_1 + v_1^2 v_3) + q_9 V_{ac} \cos \Omega t
\end{aligned}
\tag{27}
$$

where $D_n = \partial / \partial T_n$.

The solution of Equation (23) can be expressed as

$$
v_1 = A(T_2, T_4) \exp(i \omega_{1d} T_0) + \overline{A}(T_2, T_4) \exp(-i \omega_{1d} T_0)
\tag{28}
$$

where the overbar indicates the complex conjugate. The function $A(T_2, T_4)$ can be determined by eliminating the secular terms from Equations (24)–(27).

Expressing the amplitude A in the polar form $A(T_2, T_4) = \frac{1}{2} a e^{i\theta}$, where a and θ are real functions of T_2 and T_4 and separating secular terms into its real and imaginary parts, we can obtain the following average equation

$$
a' = -\frac{1}{2} c_n a + \frac{1}{2} \frac{q_9 V_{ac}}{\omega_{1d}} \sin \varphi
\tag{29}
$$

$$a\varphi' = \sigma a + \frac{1}{8}\frac{\kappa_1 a^3}{\omega_{1d}} + \frac{1}{32}\frac{\kappa_2 a^5}{\omega_{1d}} + \frac{1}{2}\frac{q_9 V_{ac}}{\omega_{1d}}\cos\varphi \tag{30}$$

where $\varphi = \sigma T_4 - \theta$, $a' = \partial a/\partial T_2 + \partial a/\partial T_4$, $\varphi' = \partial\varphi/\partial T_2 + \partial\varphi/\partial T_4$, $\kappa_1 = (q_5 - q_1\omega_{1d}^2)(\frac{10q_5}{3\omega_{1d}^2} - \frac{1}{3}q_1) + 3(q_6 - q_2\omega_{1d}^2)$ and κ_2 is expressed as Equations (A26)–(A29) in Appendix A.

Then, the steady-state frequency response can be obtained by solving the following frequency response equation

$$[(\frac{1}{2}c_n)^2 + (\sigma + \frac{1}{8}\frac{\kappa_1 a^2}{\omega_{1d}} + \frac{1}{32}\frac{\kappa_2 a^4}{\omega_{1d}})^2]a^2 = \frac{q_9^2 V_{ac}^2}{4\omega_{1d}^2} \tag{31}$$

The relationship between the resonance frequency shift and the maximum amplitude of oscillation is derived as

$$\sigma = -\frac{1}{8}\frac{\kappa_1 a_{\max}^2}{\omega_{1d}} - \frac{1}{32}\frac{\kappa_2 a_{\max}^4}{\omega_{1d}} \tag{32}$$

where $a_{\max} = q_9 V_{ac}/\omega_{1d}c_n$.

By inspection, when the $\kappa_1 a^2/8 + \kappa_2 a^4/32$ monotonically changes with the increase of amplitude, it is clear that the device will experience hardening if $\kappa_1 a^2/8 + \kappa_2 a^4/32 < 0$ and will experience softening if $\kappa_1 a^2/8 + \kappa_2 a^4/32 > 0$. If the change of the value of $\kappa_1 a^2/8 + \kappa_2 a^4/32$ is not monotonous, the system will experience hardening and softening simultaneously. For small amplitude oscillation, the hardening and softening of the system are decided by κ_1. However, with the increase of amplitude, the influence of high-order nonlinear terms on the system becomes more and more important. So, hardening and softening properties become very complex under the large amplitude vibration. Zhang et al. [22] and Nayfeh et al. [23] studied spring softening and hardening based on the traditional single degree of freedom model. However, with the increase of amplitude, it cannot describe dynamic behaviors of the system. Here, the monostable large amplitude vibration is studied with the improved single degree of freedom model. Following, taking σ as the bifurcation parameter and taking V_{dc} and V_{ac} as the unfolding parameters, we calculate unfolding of Equation (31). The traditional calculation method of the transition set of the system will lead to nonlinear equations containing high-order terms. Here, we only need to obtain parameter space of the monostable vibration. So, a new way is given.

The amplitude frequency curve can be decomposed into two parts, as the following

$$\sigma = f_1(a) = -\frac{1}{8}\frac{\kappa_1 a^2}{\omega_{1d}} - \frac{1}{32}\frac{\kappa_2 a^4}{\omega_{1d}} + \sqrt{\frac{q_9^2 V_{ac}^2}{4\omega_{1d}^2 a^2} - (\frac{1}{2}c_n)^2} \tag{33}$$

$$\sigma = f_2(a) = -\frac{1}{8}\frac{\kappa_1 a^2}{\omega_{1d}} - \frac{1}{32}\frac{\kappa_2 a^4}{\omega_{1d}} - \sqrt{\frac{q_9^2 V_{ac}^2}{4\omega_{1d}^2 a^2} - (\frac{1}{2}c_n)^2} \tag{34}$$

The monostable vibration appears when the frequency σ of the left amplitude frequency curve f_2 monotonically increases with the increase of amplitude and the frequency σ of the right amplitude frequency curve f_1 monotonically decreases with the increase of amplitude. If the curve f_1 monotonically decreases and the curve f_2 cannot monotonically change, the softening appears. On the contrary, if the curve f_2 monotonically increases and the curve f_1 cannot monotonically change, the hardening appears. If both the curve f_2 and the curve f_1 cannot monotonically change, the system will experience hardening and softening simultaneously. Here, the dynamic curve is in contact with mathematical function, which simplifies the calculation. Now, the necessary and sufficient conditions of the existence of the monostable vibration are obtained.

$$\frac{df_1(a)}{da} < 0 \text{ and } \frac{df_2(a)}{da} > 0 \text{ for } a \in [0, a_{\max}] \tag{35}$$

4. Results and Discussion

In this section, Differential Quadrature method and Finite Element method are introduced to verify the accuracy of the model. Meanwhile, with unfolding analysis and optimization theory, parameter space and maximum amplitude of the monostable vibration are obtained.

4.1. Convergence Analysis

Convergence is a key problem. With the traditional single degree of freedom model, static displacement curve cannot be obtained accurately. To solve the problem, higher dimensional model was introduced in previous studies [39]. Here, the novel single degree of freedom model is used to handle this convergence problem. To validate accuracy of our model, finite element results are obtained from the Multiphysics simulation software. Meanwhile, compared with single degree of freedom model results obtained by using Linear Galerkin method, the ascendency of the Nonlinear Galerkin method appears. Here, we consider a microbeam without scale effect and axial stress.

In Figure 2, the calculated static deflections of the microbeam obtained by using the Nonlinear Galerkin method are compared with those obtained by using Linear Galerkin method and Finite Element method. Results are presented from 0 V to pull-in voltage. It is noted from Figure 2 that the pull-in voltage predicted by the Nonlinear Galerkin method is more accurate than that predicted by Linear Galerkin method. The midpoint deflections predicted by those three methods are very close away from pull-in voltage but the midpoint deflections predicted by Linear Galerkin method deviate increasingly as pull-in is approached. The ascendency of the Nonlinear Galerkin method appears. Then, natural frequency under different DC voltage is obtained by using those three methods. As shown in Table 2, the error of results obtained by the Nonlinear Galerkin method is less than that obtained by the Linear Galerkin method. As pull-in is approached, the error of results obtained by Linear Galerkin method reaches to 9.0%. However, the error of results obtained by the Nonlinear Galerkin method is about 2.3%. When DC voltage is away from pull-in. the error caused by the Nonlinear Galerkin method is less than 1%. It is worth noting that the error is obtained by comparing the result of the theory and that of the simulation.

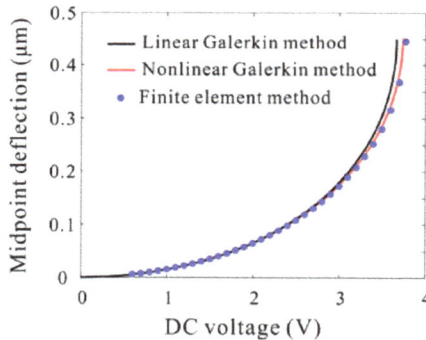

Figure 2. Comparison of the calculated midpoint deflection using nonlinear Galerkin method, linear Galerkin method and finite element method for various values of DC voltages under $d = 1\,\mu m$.

Besides, pull-in voltage is obtained by using those three methods under different gap width. As shown in Table 3, it can be seen that the result obtained by the Nonlinear Galerkin method agrees well with that obtained by Finite Element method, which demonstrates that the present analytical method is effective. However, the error of results obtained by Linear Galerkin method reaches to 7.1%.

Thus, our model is superior to the traditional single degree of freedom model. It can predict static displacement, natural frequency and pull-in voltage more accurately and convergence problem is also solved by our model.

Table 2. Natural frequency under different DC voltage when $d = 1\,\mu m$.

Case	DC Voltage (V)	Linear Galerkin Method Results (kHz)	Nonlinear Galerkin Method Results (kHz)	Finite Element Results (kHz)	Error
1	2	62.44	62.77	62.82	0.6%; 0.1%
2	2.5	59.85	60.42	60.54	1.1%; 0.2%
3	3	55.53	56.55	56.87	2.4%; 0.6%
4	3.5	44.33	47.59	48.71	9.0%; 2.3%

Table 3. Pull-in voltages under different gap width.

Case	Gap Width (μm)	Linear Galerkin Method Results (V)	Nonlinear Galerkin Method Results (V)	Finite Element Results (V)	Error
1	0.5	1.25	1.27	1.27	1.6%; 0%
2	1	3.66	3.74	3.76	2.7%; 0.5%
3	1.5	7.16	7.39	7.50	4.5%; 1.5%
4	2	11.94	12.47	12.85	7.1%; 3.0%

4.2. Static Analysis

In this section, scale effect and axial stress are considered. The static deflection and static pull-in voltage of microbeam are calculated with our model. Meanwhile, Differential Quadrature method is introduced to handle Equation (5) for numerical verification. The calculated static deflections of the microbeam with $\eta = 0.25$, $d = 1\,\mu m$ and subject to a stretched axial stress $N = 6$ are shown in Figure 3. It can be noted from this figure that the pull-in voltage predicted by the Nonlinear Galerkin method is accurate while that predicted by the Linear Galerkin method has significant error. The lower branches predicted by those three methods are very close away from pull-in voltage but the branch predicted by the Linear Galerkin method deviates increasingly as pull-in is approached. Generally, the results obtained by using the Nonlinear Galerkin method are in excellent agreement with those obtained with the Differential Quadrature method. However, the static deflections of microbeam obtained by using the Linear Galerkin method are in poor agreement with them. Specially, when DC voltage approaches zero, the upper branch predicted by Linear Galerkin method is non-convergent. The upper branch represents potential barrier of the system. When the vibration amplitude approaches potential barrier, the results predicted by Linear Galerkin method have serious errors. With the Nonlinear Galerkin method, the misconvergence of potential barrier is solved.

What's more, the calculation formula of the pull-in voltage can be obtained easily.

From Equation (19), the relationship between static displacement u_{1s} and DC voltage V_{dc} is obtained

$$u_{1s} = f(V_{dc}) \tag{36}$$

When pull-in occurs, both branches collide and destroy each other with one eigenvalue tending to zero. Thus, the pull-in voltage corresponds to a saddle-node bifurcation. So, the pull-in occurs when $dV/du_{1s} = 0$. As shown in Figure 4, the pull-in voltages under different size parameters and axial stress are given. To describe qualitatively the change of pull-in voltages, the electrostatic force coefficient α_2 should remain constant with the increase of the ratio coefficient α_1. It is noted from Figure 4 that the pull-in voltage increases with the increase of the ratio coefficient of the gap width to the mircobeam thickness. Meanwhile, a stretched axial stress can increase pull-in voltage. Driven by the same DC voltage, increasing the ratio of the gap width to the mircobeam thickness and positive axial stress is useful to prevent pull-in.

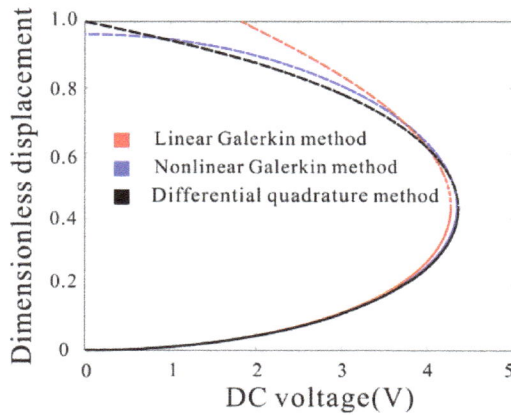

Figure 3. Comparison of the calculated maximum non-dimensional deflection using Nonlinear Galerkin method, Differential Quadrature method and linear Galerkin method for various values of DC voltages (Solid line: stable; dashed line: unstable).

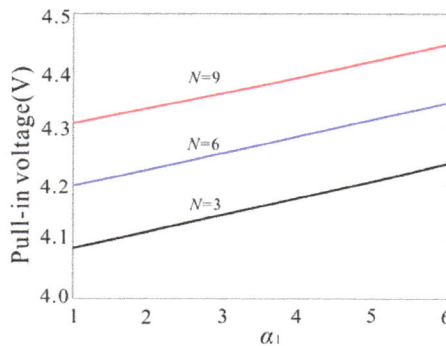

Figure 4. The influence of ratio of the gap width to mircobeam thickness and axial stress on the pull-in voltage.

4.3. Dynamic Analysis

The monostable vibration is desired for many applications, such as microbeam resonator [40]. Here, parameter space and maximum amplitude about the monostable vibration are obtained.

4.3.1. Small Vibration

In Figure 5, we study the influence of the ratio of the gap width to the mircobeam thickness and DC voltage on the hardening and softening properties of the system under the small amplitude oscillation. It is noted that the increase of the DC voltage and the decrease of the gap width can lead to softening phenomenon with $\kappa_1 > 0$. On the contrary, the decrease of the voltage and the increase of the gap width can lead to hardening phenomenon with $\kappa_1 < 0$. It is found that the mechanical spring is responsible for the hardening behavior and the electrostatic force is responsible for the softening behavior. From Figure 5, the curve represents the boundary between the softening area and hardening area. Near the boundary, there is no softening phenomenon or hardening phenomenon and the system will experience monostable vibration.

Three types of parameters (point A, point B and point C) are taken from softening area, hardening area and boundary as shown in Figure 5. And the amplitude frequency response curves of them are given as shown in Figure 6.

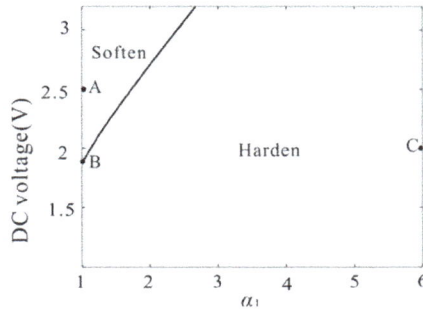

Figure 5. The parameter space of softening and hardening under the small vibration amplitudes.

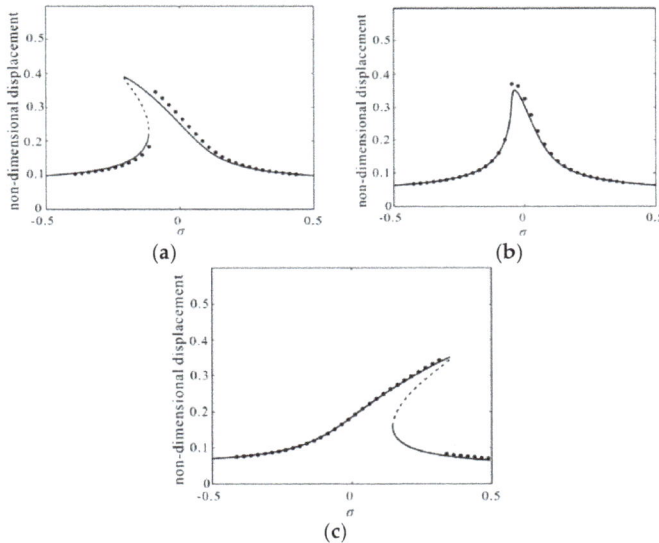

Figure 6. (**a–c**) Comparison of the frequency response curve obtained by nonlinear Galerkin method (line) and differential quadrature method (dotted line) corresponding to A–C in Figure 5 (solid line: stable; dashed line: unstable).

Figure 6a shows a representative frequency response to our problem when $\kappa_1 > 0$ in the case of $V_{dc} = 2.5$, $V_{ac} = 0.015$, $\alpha_1 = 1$. Here, appropriate excitation voltages are needed to introduce softening nonlinearity and prevent pull-in. And their stability is studied by using Routh Criterion. Figure 6b shows monostable vibration when $\kappa_1 = 0$ in the case of $V_{dc} = 1.86$, $V_{ac} = 0.022$, $\alpha_1 = 1$. At this time, the DC voltage and gap width should satisfy certain relations as shown in Figure 5. Figure 6c shows hardening nonlinearity when $\kappa_1 < 0$ in the case of $V_{dc} = 2$, $V_{ac} = 0.02$, $\alpha_1 = 6$. Meanwhile, the results obtained by the Nonlinear Galerkin method are compared with them obtained by Differential Quadrature method. Here, we apply the way of frequency sweep into the Differential Quadrature method and they are in excellent agreement with each other.

4.3.2. Monostable Large Amplitude Vibration

Due to the existence of nonlinear electrostatic force and geometric nonlinearity, it is hard to realize the monostable large amplitude vibration. The traditional single degree of freedom model is not enough to characterize the frequency response under the large amplitude vibration. In this section, we try to qualitatively study the monostable large amplitude vibration with the improved single degree of freedom model. To verify the validity of the results, the Differential Quadrature method is used to compare with the Nonlinear Galerkin method.

Section 4.3.1 shows that the nonlinear electrostatic force will make the device experience softening and the geometric nonlinearity will make device experience hardening. Under the small amplitude oscillation, to realize the monostable vibration, parameters should be selected near the boundary in Figure 5. And, from Equation (5), it is found that the growth rate of nonlinear electrostatic force is faster than that of geometric nonlinearity with the increase of amplitude of oscillation. The softening becomes dominant when the gap width decreases. So, under the large amplitude oscillation, the device tends to experience softening. In order to realize the monostable large amplitude vibration, we need select parameters near the boundary in Figure 5 and control softening with the increase of amplitude. Besides, as the AC voltage increases, the amplitude increases. More solutions will appear only when amplitude increases to a critical value.

Then, from Equation (35), parameter space of the monostable vibration with the different specification will be studied. As shown in Figure 7, the boundary between monostable vibration and multistable vibration in the case of $\eta = 0.25$ and $\eta = 0.5$ is given. When AC voltage is less than that of the boundary, monostable vibration appears. It is found that there is only one solution when the DC voltage or AC voltage is very small, which can be proved with Equation (31). With a small vibration force, the system is equivalent to linear vibration. It is impossible for the system to generate more than one solution. With the increase of vibration force, to obtain the monostable vibration, strict parameter conditions are given. It is noted that there is one peak under the different specification, where the exciting force is put to the maximum. Under the monostable vibration, the system becomes almost linear. And the amplitude of linear vibration is proportional to the exciting force. Then, the maximum amplitude is obtained near the peak in the Figure 7a. What's more, Figure 7 shows the parameter space of the monostable vibration increases with the increase of the scale effect.

Besides, the AC voltage of the peak decreases as the ratio of the gap width to the mircobeam thickness increases, which can be explained with Figure 5. With the increase of the ratio of the gap width to the thickness of the mircobeam, a relatively large DC voltage is required to counteract hardening of the system. However, as the amplitude increases, the softening is more and more obvious. So, a relatively small AC voltage is needed to prevent bifurcation. On the contrary, if the ratio of the gap width to the mircobeam thickness is too small, a small enough DC voltage is required to counteract hardening. Meanwhile, a relatively large AC voltage is required to produce large amplitude. However, the condition $V_{dc} >> V_{ac}$ is false. The second-order item of the AC voltage cannot be ignored, which can lead to multiple frequency vibration.

The ratio of the gap width to the mircobeam thickness is the key to realize monostable large amplitude vibration and it decides the maximum amplitude that can be realized with appropriate DC and AC voltage. Reasonable decreasing the ratio of the gap width to the mircobeam thickness, the system can realize monostable large amplitude vibration. As we know, monostable large amplitude vibration is desired for many applications, such as microbeam resonator, which can eliminate the dynamic bifurcation phenomenon and increase the vibration energy.

To verify the validity of the results, the Differential Quadrature method is proposed to compare with the Nonlinear Galerkin method. Figure 8a–f show the frequency response corresponding to A–F shown in Figure 7.

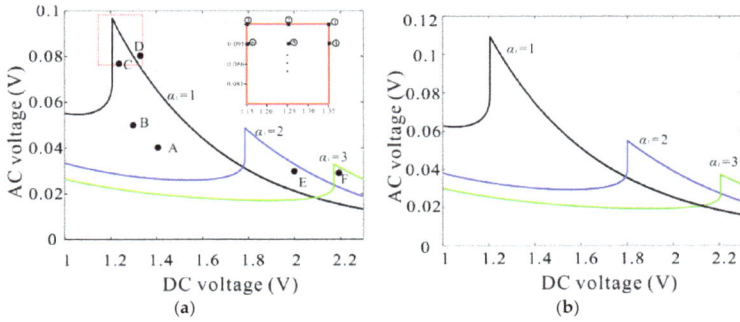

Figure 7. The $V_{ac} - V_{dc}$ parameter space under different ratio of the gap width and the thickness of the mircobeam in the case of $\eta = 0.25$ (**a**) and $\eta = 0.5$ (**b**) (The parameter area below the curve is monostable parameter space).

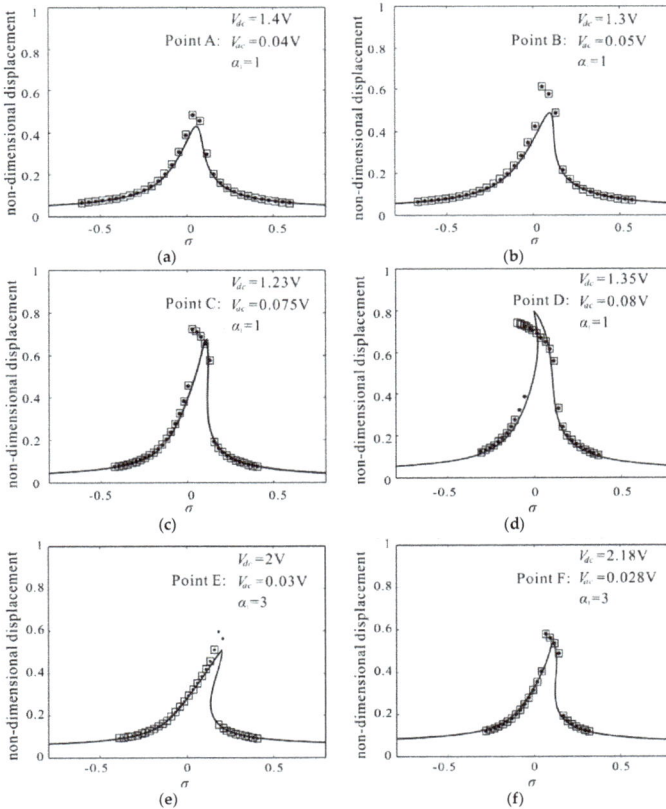

Figure 8. Comparison of the frequency response curve obtained by Nonlinear Galerkin method (solid line) and Differential Quadrature method: (**a**–**f**) corresponding to A–F in Figure 7a (dotted line represents result obtained by sweeping up the frequency; rectangle represents result obtained by sweeping down the frequency).

Here, the frequency response is calculated with sweeping up the frequency and sweeping down the frequency by Differential Quadrature method. As shown in Figure 8a–c, f, the results obtained by sweep-up case and sweep-down case are consistent, which conforms to the monostable vibration shown in Figure 7. From Figure 7, the voltage of point D exceeds parameter space of monostable vibration, which leads to spring softening dominance as shown in Figure 8d. Meanwhile, the DC voltage of point E is relatively small, which leads to spring hardening dominance as shown in Figure 8e. From Figure 8, with the increase of the amplitude, the deviation between the results obtained by the Nonlinear Galerkin method and those obtained by Differential Quadrature method becomes more and more obvious. But the error cannot affect our conclusion that the improved single degree of freedom model can study qualitatively monostable large amplitude vibration. To quantitatively study monostable large amplitude vibration, an optimization theory is proposed.

The formula of maximum amplitude $q_9 V_{ac}/\omega_{1d} c_{nd}$ is used near the peak shown in Figure 7. Then, the maximum amplitude of monostable vibration under different ratio of the gap width to the mircobeam thickness and size effect is predicted. And, an optimization theory, which is based on the following three points, is proposed to verify theoretical prediction results.

(a) Under monostable vibration, κ_1 and κ_2 approximate to zero. The system is equivalent to approximate linear vibration.

(b) The maximum amplitude of approximate linear vibration is proportional to exciting force that is decided with the product of DC voltage and AC voltage.

(c) Optimization parameters, which can realize the maximum amplitude of monostable vibration, are taken near the peak regions in Figure 7 (for example, the red frame with $\alpha_1 = 1$).

We take a rectangular area near the peak region and discretize it into many points as shown in Figure 7. As serial number increases, exciting force decreases. Differential Quadrature method is used to calculate the frequency response according to the order. When results obtained by sweeping up the frequency and sweeping down the frequency are consistent, the calculation stops and the maximum amplitude of monostable vibration is obtained as shown in Figure 9. It is found that the maximum amplitude increases with the decrease of the ratio of the gap width to the mircobeam thickness. Meanwhile, as size effect increases, the maximum amplitude increases. The results obtained by theoretical prediction are qualitative agreement with them obtained by Differential Quadrature method. However, theoretical results are larger than numerical ones. With the decrease of the ratio of gap width to thickness of mircobeam, the deviation between theoretical prediction and numerical result becomes more and more obvious. In the next study, in order to improve the calculation accuracy, more space dimensions should be taken in the Nonlinear Galerkin method.

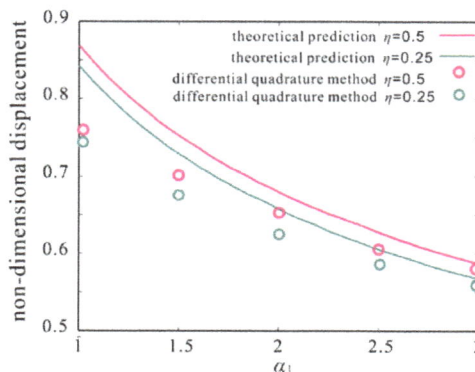

Figure 9. Comparison of the maximum amplitudes obtained by theoretical prediction and Differential Quadrature method.

In this section, the improved single degree of freedom model can describe monostable large amplitude vibration qualitatively. Although, when amplitude exceeds half of the gap, the error between reduced-order model and continuum model becomes obvious, our proposed model is significant to study the relationship between maximum amplitude and physical parameter.

5. Conclusions

With the Nonlinear Galerkin method, we propose a novel approach to generate an improved single degree of freedom model for electrically actuated microbeam-based MEMS and use it to study the static and dynamic behaviors of these devices. Specially, the monostable vibration is theoretically investigated in size effected MEMS via the low dimensional model. The proposed theoretical results maintain a good situation consistency with the results obtained by Differential Quadrature method. Besides, the Finite element results of case studies are used to verify the accuracy of the model. Compared with the results obtained by the Linear Galerkin method, the model has obvious superiority.

What's more, the monostable large amplitude vibration eliminates dynamic bifurcation phenomenon, improves the stability of the system and increases the vibration energy, which is desired for many applications. Parameter space and maximum amplitude of the monostable vibration are obtained by using unfolding analysis and optimization theory for the first time. It is found that reasonable decreasing the ratio of the gap width to the thickness of the mircobeam is the key to realize monostable large amplitude vibration. Besides, the Nonlinear Galerkin method gives a way to convert partial differential equation into ordinary differential equation.

Acknowledgments: The work was supported by the National Natural Science Foundation of China (Grant Nos. 11372210 and 11702192).

Author Contributions: Lei Li and Qichang Zhang conceived and obtained the low dimensional model; Lei Li and Wei Wang contributed theoretical analysis; Lei Li, Wei Wang, and Jianxin Han analyzed the data; Lei Li wrote the paper.

Conflicts of Interest: The authors declare no conflict of interest.

Appendix A

Equations (A1)–(A15) represent the coefficients produced in the Nonlinear Galerkin process; Equations (A16)–(A25) represent the coefficients that are obtained by eliminating static displacement; Equations (A26)–(A29) represent the coefficients produced in the perturbation process.

$$n_0 = 1 - 2\frac{\alpha_2 V_{dc}^2 \int_0^1 \phi_1^2 \phi_3 dx \int_0^1 \phi_3 dx}{\omega_3^2} \tag{A1}$$

$$n_1 = -2\int_0^1 \phi_1^3 dx + 2\frac{\alpha_2 V_{dc}^2}{\omega_3^2}\int_0^1 \phi_1^3 \phi_3 dx \int_0^1 \phi_3 dx \tag{A2}$$

$$n_2 = \int_0^1 \phi_1^4 dx - 4\left(\int_0^1 \phi_1^2 \phi_3 dx\right)^2 \frac{\omega_1^2}{\omega_3^2} \tag{A3}$$

$$n_3 = 6\frac{\omega_1^2}{\omega_3^2}\int_0^1 \phi_1^2 \phi_3 dx \int_0^1 \phi_3^3 \phi_3 dx - \frac{2\alpha_1}{\omega_3^2}\int_0^1 \phi_1^2 \phi_3 dx \int_0^1 \phi_1'' \phi_3 dx \int_0^1 \phi_1'^2 dx \tag{A4}$$

$$n_4 = \frac{4\alpha_1}{\omega_3^2}\int_0^1 \phi_1^2 \phi_3 dx \int_0^1 \phi_1 \phi_1'' \phi_3 dx \int_0^1 \phi_1'^2 dx - 2\frac{\omega_1^2}{\omega_3^2}\left(\int_0^1 \phi_1^3 \phi_3 dx\right)^2 + \frac{2\alpha_1}{\omega_3^2}\int_0^1 \phi_1^3 \phi_3 dx \int_0^1 \phi_1'' \phi_3 dx \int_0^1 \phi_1'^2 dx \tag{A5}$$

$$m_1 = \alpha_1 \int_0^1 \phi_1'' \phi_1 dx \int_0^1 \phi_1'^2 dx + 4\omega_3^2\left(\int_0^1 \phi_1^2 \phi_3 dx\right)^2 \frac{\omega_1^2}{\omega_3^2} \tag{A6}$$

$$m_2 = 4\left(\int_0^1 \phi_1^2 \phi_3 dx\right)^2 \tag{A7}$$

$$m_3 = 2\alpha_2 V_{dc}^2 \int_0^1 \phi_1^2 \phi_3 dx \int_0^1 \phi_3 dx \tag{A8}$$

$$m_4 = \alpha_2 V_{dc}^2 \int_0^1 \phi_1 dx \tag{A9}$$

$$m_5 = \alpha_2 \int_0^1 \phi_1 dx \tag{A10}$$

$$\chi = -\omega_3^2 \int_0^1 \phi_1^3 \phi_3 dx + 2\alpha_1 \int_0^1 \phi_1'' \phi_1 dx \int_0^1 \phi_1' \phi_3' dx + \alpha_1 \int_0^1 \phi_1^2 dx \int_0^1 \phi_1 \phi_3'' dx \tag{A11}$$

$$m_6 = \chi \frac{\alpha_2 V_{dc}^2}{\omega_3^2} \int_0^1 \phi_3 dx \tag{A12}$$

$$m_7 = 2\omega_3^2 \int_0^1 \phi_1^2 \phi_3 dx (\frac{\alpha_1}{\omega_3^2} \int_0^1 \phi_1'' \phi_3 dx \int_0^1 \phi_1'^2 dx - \frac{\omega_1^2}{\omega_3^2} \int_0^1 \phi_1^3 \phi_3 dx) + \frac{2\omega_1^2 \chi}{\omega_3^2} \int_0^1 \phi_1^2 \phi_3 dx - 2\alpha_1 \int_0^1 \phi_1'^2 dx \int_0^1 \phi_1^2 \phi_1'' dx \tag{A13}$$

$$m_8 = -2 \int_0^1 \phi_1^2 \phi_3 dx \int_0^1 \phi_1^3 \phi_3 dx + \frac{2\chi}{\omega_3^2} \int_0^1 \phi_1^2 \phi_3 dx \tag{A14}$$

$$m_9 = \alpha_1 \int_0^1 \phi_1'^2 dx \int_0^1 \phi_1^3 \phi_1'' dx - 4\alpha_1 \int_0^1 \phi_1^2 \phi_3 dx \int_0^1 \phi_1 \phi_1'' \phi_3 dx \int_0^1 \phi_1'^2 dx + \frac{\alpha_1 \chi}{\omega_3^2} \int_0^1 \phi_1^2 dx \int_0^1 \phi_1 \phi''_3 dx - \frac{\omega_1^2 \chi}{\omega_3^2} \int_0^1 \phi_1^3 \phi_3 dx \tag{A15}$$

$$q_1 = \frac{-(n_1 + 2n_2 u_{1s} + 3n_3 u_{1s}^2 + 4n_4 u_{1s}^3) + 2m_2 u_{1s} + 3m_8 u_{1s}^2}{n_0 + n_1 u_{1s} + (n_2 - m_2) u_{1s}^2 + (n_3 - m_8) u_{1s}^3 + n_4 u_{1s}^4} \tag{A16}$$

$$q_2 = \frac{m_2 - n_2 + 3(m_8 - n_3) u_{1s} - 6n_4 u_{1s}^2}{n_0 + n_1 u_{1s} + (n_2 - m_2) u_{1s}^2 + (n_3 - m_8) u_{1s}^3 + n_4 u_{1s}^4} \tag{A17}$$

$$q_3 = \frac{m_8 - n_3 - 4n_4 u_{1s}}{n_0 + n_1 u_{1s} + (n_2 - m_2) u_{1s}^2 + (n_3 - m_8) u_{1s}^3 + n_4 u_{1s}^4} \tag{A18}$$

$$q_4 = \frac{-n_4}{n_0 + n_1 u_{1s} + (n_2 - m_2) u_{1s}^2 + (n_3 - m_8) u_{1s}^3 + n_4 u_{1s}^4} \tag{A19}$$

$$q_5 = \frac{-\omega_1^2 (3n_2 u_{1s} + n_1 + 6n_3 u_{1s}^2 + 10n_4 u_{1s}^3) + m_6 + 3m_1 u_{1s} + 6m_7 u_{1s}^2 + 10m_9 u_{1s}^3}{n_0 + n_1 u_{1s} + (n_2 - m_2) u_{1s}^2 + (n_3 - m_8) u_{1s}^3 + n_4 u_{1s}^4} \tag{A20}$$

$$q_6 = \frac{-\omega_1^2 (n_2 + 4n_3 u_{1s} + 10n_4 u_{1s}^2) + m_1 + 4m_7 u_{1s} + 10m_9 u_{1s}^2}{n_0 + n_1 u_{1s} + (n_2 - m_2) u_{1s}^2 + (n_3 - m_8) u_{1s}^3 + n_4 u_{1s}^4} \tag{A21}$$

$$q_7 = \frac{-\omega_1^2 (n_3 + 5n_4 u_{1s}) + m_7 + 5m_9 u_{1s}}{n_0 + n_1 u_{1s} + (n_2 - m_2) u_{1s}^2 + (n_3 - m_8) u_{1s}^3 + n_4 u_{1s}^4} \tag{A22}$$

$$q_8 = \frac{m_9 - \omega_1^2 n_4}{n_0 + n_1 u_{1s} + (n_2 - m_2) u_{1s}^2 + (n_3 - m_8) u_{1s}^3 + n_4 u_{1s}^4} \tag{A23}$$

$$q_9 = \frac{2m_5 V_{dc}}{n_0 + n_1 u_{1s} + (n_2 - m_2) u_{1s}^2 + (n_3 - m_8) u_{1s}^3 + n_4 u_{1s}^4} \tag{A24}$$

$$\omega_{1d}^2 = \frac{(n_0 + 2n_1 u_{1s} + 3n_2 u_{1s}^2 + 4n_3 u_{1s}^3 + 5n_4 u_{1s}^4) \omega_1^2 - 5m_9 u_{1s}^4 - 4m_7 u_{1s}^3 - 3m_1 u_{1s}^2 - 2m_6 u_{1s} - m_3}{n_0 + n_1 u_{1s} + (n_2 - m_2) u_{1s}^2 + (n_3 - m_8) u_{1s}^3 + n_4 u_{1s}^4} \tag{A25}$$

$$\chi_1 = q_2 \omega_{1d}^2 - q_6 + \frac{q_1 \omega_{1d}^2 - q_5}{3\omega_{1d}^2} (5q_1 \omega_{1d}^2 - 2q_5) \tag{A26}$$

$$\chi_2 = \frac{q_1 \omega_{1d}^2 - q_5}{\omega_{1d}^2} [q_6 + 2q_2 \omega_{1d}^2 + \frac{4}{3} \kappa_1 + \frac{q_1 \omega_{1d}^2 - q_5}{3\omega_{1d}^2} (4q_5 - 8q_1)] - 4q_7 + 4q_3 \omega^2 - q_1 \kappa_1 + \frac{9q_1 \omega_{1d}^2 - 2q_5}{8\omega_{1d}^2} \chi_1 \tag{A27}$$

$$\chi_3 = \frac{q_1 \omega_{1d}^2 - q_5}{9\omega_{1d}^2} (19q_5 - 4q_1 \omega_{1d}^2 - 45q_6 + 6q_2 \omega_{1d}^2) + 3q_7 - 3q_3 \omega_{1d}^2 + q_1 \kappa_1 \tag{A28}$$

$$\begin{aligned}
\kappa_2 &= \left(\frac{\chi_2}{3\omega_{1d}^2} + \frac{2\chi_3}{\omega_{1d}^4}\right)(2q_5 - 5q_1\omega_{1d}^2) + \chi_1\left[\frac{3q_6 - 11q_2\omega_{1d}^2}{8\omega_{1d}^2} + \frac{q_1\omega_{1d}^2 - q_5}{\omega_{1d}^2}\left(\frac{q_5}{12\omega_{1d}^2} - \frac{q_1}{2}\right)\right] \\
&\quad + \frac{q_1\omega_{1d}^2 - q_5}{\omega_{1d}^2}\left[\frac{q_1\omega_{1d}^2 - q_5}{\omega_{1d}^2}\left(\frac{16q_6}{3} - \frac{8q_2\omega_{1d}^2}{3}\right) - \frac{56}{3}q_7 + \frac{26}{3}q_3\omega_{1d}^2\right] + \kappa_1\left(\frac{q_5 - q_1\omega_{1d}^2}{3\omega_{1d}^2}q_1 + 3q_2\right) \\
&\quad + 10q_8 - 10\omega_{1d}^2 q_4
\end{aligned} \tag{A29}$$

References

1. Kouravand, S. Design and modeling of some sensing and actuating mechanisms for MEMS applications. *Appl. Math. Model.* **2011**, *35*, 5173–5181. [CrossRef]
2. Rhoads, J.F.; Shaw, S.W.; Turner, K.L. Nonlinear Dynamics and Its Applications in Micro- and Nanoresonators. *J. Dyn. Syst. Meas. Control* **2010**, *132*, 034001. [CrossRef]
3. Zhang, Z.; Liang, J.; Zhang, D.; Pang, W.; Zhang, H. A Novel Bulk Acoustic Wave Resonator for Filters and Sensors Applications. *Micromachines* **2015**, *6*, 1306–1316. [CrossRef]
4. Jung, J.; Kim, P.; Lee, J.I.; Seok, J. Nonlinear dynamic and energetic characteristics of piezoelectric energy harvester with two rotatable external magnets. *Int. J. Mech. Sci.* **2015**, *92*, 206–222. [CrossRef]
5. Lee, J.; Jeong, B.; Park, S.; Eun, Y.; Kim, J. Micromachined Resonant Frequency Tuning Unit for Torsional Resonator. *Micromachines* **2017**, *8*, 342. [CrossRef]
6. Ramanan, A.; Teoh, Y.; Ma, W.; Ye, W. Characterization of a Laterally Oscillating Microresonator Operating in the Nonlinear Region. *Micromachines* **2016**, *7*, 132. [CrossRef]
7. Toan, N.; Shimazaki, T.; Inomata, N.; Song, Y.; Ono, T. Design and Fabrication of Capacitive Silicon Nanomechanical Resonators with Selective Vibration of a High-Order Mode. *Micromachines* **2017**, *8*, 312. [CrossRef]
8. Song, Z.K.; Li, H.X.; Sun, K.B. Adaptive dynamic surface control for MEMS triaxial gyroscope with nonlinear inputs. *Nonlinear Dyn.* **2014**, *78*, 173–182. [CrossRef]
9. Verotti, M.; Dochshanov, A.; Belfiore, N.P. A Comprehensive Survey on Microgrippers Design: Mechanical Structure. *J. Mech. Des.* **2017**, *139*, 060801. [CrossRef]
10. Dochshanov, A.; Verotti, M.; Belfiore, N.P. A Comprehensive Survey on Microgrippers Design: Operational Strategy. *J. Mech. Des.* **2017**, *139*, 070801. [CrossRef]
11. Rhoads, J.F.; Shaw, S.W.; Turner, K.L. The nonlinear response of resonant microbeam systems with purely-parametric electrostatic actuation. *J. Micromech. Microeng.* **2006**, *16*, 890–899. [CrossRef]
12. Zhong, Z.Y.; Zhang, W.M.; Meng, G.; Wu, J. Inclination Effects on the Frequency Tuning of Comb-Driven Resonators. *J. Microelectromechan. Syst.* **2013**, *22*, 865–875. [CrossRef]
13. Han, J.X.; Zhang, Q.C.; Wang, W. Design considerations on large amplitude vibration of a doubly clamped microresonator with two symmetrically located electrodes. *Commun. Nonlinear Sci. Numer. Simul.* **2015**, *22*, 492–510. [CrossRef]
14. Masri, K.M.; Shao, S.; Younis, M.I. Delayed feedback controller for microelectromechanical systems resonators undergoing large motion. *J. Vib. Control* **2015**, *13*, 2604–2615. [CrossRef]
15. Tadi Beni, Y.; Koochi, A.; Abadyan, M. Theoretical study of the effect of Casimir force, elastic boundary conditions and size dependency on the pull-in instability of beam-type NEMS. *Phys. E Low Dimens. Syst. Nanostruct.* **2011**, *43*, 979–988. [CrossRef]
16. Mousavi, T.; Bornassi, S.; Haddadpour, H. The effect of small scale on the pull-in instability of nano-switches using DQM. *Int. J. Solids Struct.* **2013**, *50*, 1193–1202. [CrossRef]
17. Ghayesh, M.H.; Farokhi, H.; Amabili, M. Nonlinear behaviour of electrically actuated MEMS resonators. *Int. J. Eng. Sci.* **2013**, *71*, 137–155. [CrossRef]
18. Ma, H.; Gao, X.; Reddy, J. A microstructure-dependent Timoshenko beam model based on a modified couple stress theory. *J. Mech. Phys. Solids* **2008**, *56*, 3379–3391. [CrossRef]
19. Anthoine, A. Effect of couple-stresses on the elastic bending of beams. *Int. J. Solids Struct.* **2000**, *37*, 1003–1018. [CrossRef]
20. Yang, F.; Chong, A.C.M.; Lam, D.C.C.; Tong, P. Couple stress-based strain gradient theory for elasticity. *Int. J. Solids Struct.* **2002**, *39*, 2731–2743. [CrossRef]
21. Guo, C.Z.; Gary, K.F. Behavioral Modeling of a CMOS–MEMS Nonlinear Parametric Resonator. *J. Microelectromech. Syst.* **2013**, *22*, 1447–1459. [CrossRef]

22. Zhang, W.M.; Meng, G. Nonlinear Dynamic Analysis of Electrostatically Actuated Resonant MEMS Sensors under Parametric Excitation. *IEEE Sens. J.* **2007**, *7*, 370–380. [CrossRef]

23. Nayfeh, A.H.; Lacarbonara, W. On the Discretization of Distributed-Parameter Systems with Quadratic and Cubic Nonlinearities. *Nonlinear Dyn.* **1997**, *13*, 203–220. [CrossRef]

24. Zhang, W.M.; Meng, G. Nonlinear dynamical system of micro-cantilever under combined parametric and forcing excitations in MEMS. *Sens. Actuators A Phys.* **2005**, *119*, 291–299. [CrossRef]

25. Younis, M.I.; Abdel-Rahman, E.M.; Nayfeh, A.H. A Reduced-Order Model for Electrically Actuated Microbeam-Based MEMS. *J. Microelectromech. Syst.* **2003**, *12*, 672–680. [CrossRef]

26. Sadeghian, H.; Rezazadeh, G.; Peter, M.O. Application of the Generalized Differential Quadrature Method to the Study of Pull-In Phenomena of MEMS Switches. *J. Microelectromech. Syst.* **2007**, *16*, 1334–1340. [CrossRef]

27. Sadeghian, H.; Rezazadeh, G. Comparison of generalized differential quadrature and Galerkin methods for the analysis of micro-electro-mechanical coupled systems. *Commun. Nonlinear Sci. Numer. Simul.* **2009**, *14*, 2807–2816. [CrossRef]

28. Ilyas, S.; Ramini, A.; Arevalo, A.; Younis, M.I. An Experimental and Theoretical Investigation of a Micromirror under Mixed-Frequency Excitation. *J. Microelectromech. Syst.* **2015**, *24*, 1124–1131. [CrossRef]

29. Younis, M.I.; Ouakad, H.M.; Alsaleem, F.M.; Miles, R.; Cui, W. Nonlinear Dynamics of MEMS Arches under Harmonic Electrostatic Actuation. *J. Microelectromech. Syst.* **2010**, *19*, 647–656. [CrossRef]

30. Younis, M.I.; Nayfeh, A.H. A study of the nonlinear response of a resonant microbeam to an electric actuation. *Nonlinear Dyn.* **2003**, *31*, 91–117. [CrossRef]

31. Alcheikh, N.; Ramini, A.; Abdullah, M.; Younis, M.I. Tunable Clamped Guided Arch Resonators Using Electrostatically Induced Axial Loads. *Micromachines* **2017**, *8*, 14. [CrossRef]

32. Masri, K.M.; Younis, M.I. Investigation of the dynamics of a clamped–clamped microbeam near symmetric higher order modes using partial electrodes. *Int. J. Dyn. Control* **2015**, *3*, 173–182. [CrossRef]

33. Nayfeh, A.H.; Younis, M.I.; Abdel-Rahman, E.M. Dynamic pull-in phenomenon in MEMS resonators. *Nonlinear Dyn.* **2006**, *48*, 153–163. [CrossRef]

34. Ouakad, H.M.; Younis, M.I. The dynamic behavior of MEMS arch resonators actuated electrically. *Int. J. Nonlinear Mech.* **2010**, *45*, 704–713. [CrossRef]

35. Cao, D.Q. A novel order reduction method for nonlinear dynamical system under external periodic excitations. *Sci. China Technol. Sci.* **2010**, *53*, 684–691. [CrossRef]

36. Matthies, H.G.; Meyer, M. Nonlinear Galerkin methods for the model reduction of nonlinear dynamical systems. *Comput. Struct.* **2003**, *81*, 1277–1286. [CrossRef]

37. Kang, W.; Zhang, J.Z.; Ren, S.; Lei, P.F. Nonlinear Galerkin method for low-dimensional modeling of fluid dynamic system using POD modes. *Commun. Nonlinear Sci. Numer. Simul.* **2015**, *22*, 943–952. [CrossRef]

38. Zhang, W.; Yan, H.; Peng, Z.; Meng, G. Electrostatic pull-in instability in MEMS/NEMS, A review. *Sens. Actuators A Phys.* **2014**, *214*, 187–218. [CrossRef]

39. Abdel-Rahman, E.M.; Younis, M.I.; Nayfeh, A.H. Characterization of the mechanical behavior of an electrically actuated microbeam. *J. Micromech. Microeng.* **2002**, *12*, 759–766. [CrossRef]

40. Zhang, W.; Hu, K.; Peng, Z.; Meng, G. Tunable micro- and nanomechanical resonators. *Sensors* **2015**, *15*, 26478–26566. [CrossRef] [PubMed]

micromachines

MDPI

Article

Acceleration Sensitivity in Bulk-Extensional Mode, Silicon-Based MEMS Oscillators

Beheshte Khazaeili [1,*], Jonathan Gonzales [2] and Reza Abdolvand [1]

[1] Department of Electrical and Computer Engineering, University of Central Florida, Orlando, FL 32816, USA; reza@eecs.ucf.edu

[2] School of Electrical and Computer Engineering, Oklahoma State University, Tulsa, OK 74074, USA; Jonathan.gonzales@okstate.edu

* Correspondence: beheshteh.khazaeili@knights.ucf.edu; Tel.: +1-407-823-1760

Received: 31 March 2018; Accepted: 9 May 2018; Published: 12 May 2018

Abstract: Acceleration sensitivity in silicon bulk-extensional mode oscillators is studied in this work, and a correlation between the resonator alignment to different crystalline planes of silicon and the observed acceleration sensitivity is established. It is shown that the oscillator sensitivity to the applied vibration is significantly lower when the silicon-based lateral-extensional mode resonator is aligned to the <110> plane compared to when the same resonator is aligned to <100>. A finite element model is developed that is capable of predicting the resonance frequency variation when a distributed load (i.e., acceleration) is applied to the resonator. Using this model, the orientation-dependent nature of acceleration sensitivity is confirmed, and the effect of material nonlinearity on the acceleration sensitivity is also verified. A thin-film piezoelectric-on-substrate platform is chosen for the implementation of resonators. Approximately, one order of magnitude higher acceleration sensitivity is measured for oscillators built with a resonator aligned to the <100> plane versus those with a resonator aligned to the <110> plane (an average of ~5.66 × 10^{-8} (1/g) vs. ~3.66 × 10^{-9} (1/g), respectively, for resonators on a degenerately n-type doped silicon layer).

Keywords: MEMS resonators; acceleration sensitivity; vibration sensitivity; nonlinearity

1. Introduction

In recent years, the application of micro-machined, silicon-based resonators in timing has been growing steadily [1–3]. Internet of things, mobile and wearable, automotive, and smart infrastructure monitoring are some examples of applications where extremely small and ultra-stable MEMS-based oscillators could play an important role in the system performance. The stability of these oscillators is affected by environmental conditions such as temperature, humidity, pressure, magnetic field, acceleration/vibration, etc. A change in one or more of these environmental conditions can either vary the resonance frequency of the resonator or the oscillator loop phase [4]. In the former case, any change in the resonance frequency of the resonator will directly impact the oscillator frequency, as other electronic components are assumed to be broadband. In the latter case, change in the phase of the oscillator loop is compensated for by a shift in the oscillation frequency, so that the loop phase is maintained at 360°, which is a required condition for oscillation [4].

A significant mechanical vibration could exist in the typical operating environment of the electronics utilized in certain applications such as cell-phone towers, aircrafts, automotives, aerospace vehicles, and radar towers. Such vibration results in frequency instability and performance deterioration of the oscillators onboard. Applied acceleration to the oscillator can induce a change in the resonance frequency of the mechanical resonator, as the stress applied to the resonator through the suspension tethers will change the instantaneous resonance frequency of the device. Alternatively, the stray capacitance and inductance of the circuit could change as the circuit board is deformed due to the applied vibration, resulting in variation of the oscillator loop phase [4].

In order to characterize the shift in resonance frequency of a mechanical resonator due to the applied stress, a parameter defined as a stress-frequency coefficient is typically used. The stress-frequency coefficient is shown to depend on the combination of two factors: the geometrical deformation and the nonlinear elastic properties of the material used in the resonator [5]. In quartz crystal resonators, this coefficient is extensively studied and proven to be dependent on the crystal cut [6–11]. Several methods have been used for reducing the vibration sensitivity of quartz-based oscillators. For example, it is suggested that compensation of the acceleration-induced shift in oscillation frequency is possible by applying an appropriately phase-shifted electrical signal to the crystal's electrodes that is proportional to the mechanical vibration [12]. Active compensation of acceleration sensitivity has also been studied in recent years. In [13–15], authors have achieved low vibration-sensitive oscillators through electrical compensation of the oscillator output using sensors that are strategically mounted to accurately measure the applied acceleration. A similar method of electronic vibration-induced noise cancelation has been used for optoelectronic oscillators as well [16]. In a different approach, two crystals with anti-parallel acceleration-sensitivity characteristics have been used in an oscillator circuit so that the net acceleration sensitivity vector can be adjusted to zero [11,17,18]. Similar compensation methods using discrete or stacked crystals are represented in [19]. Choosing an appropriate shape and place for mounting supports in order to reduce stress, and also employing precise fabrication techniques to accurately locate and shape the designed mounting structures, are other important factors considered in previous works [20]. The type of the material used in the oscillator is also another degree of freedom that could be utilized for reducing acceleration sensitivity. As explained in [21], experimental results obtained from two different air-dielectric cavity oscillators made from two different materials, ceramic and aluminum, showed almost one sixth times lower acceleration sensitivity for the one made of ceramic. Optical oscillators also showed low acceleration sensitivities due to their small size [22]. Other interesting improvements are reported in the field of optical cavity resonators. The acceleration sensitivity of a cavity-stabilized laser is decreased significantly using the feedforward correction of acceleration [23]. In addition, recent research on the acceleration sensitivity of optical cavity resonators shows that performance deterioration due to the external mechanical vibration can be minimized by the appropriate design of the cavity structure [24–26].

As mentioned above, material nonlinearity is an important factor affecting the sensitivity of the resonators to mechanical vibration [4,8,27,28]. Since nonlinear elastic constants in crystalline material are anisotropic, the orientation-dependency of the vibration sensitivity due to this nonlinearity is potentially predictable. It has been shown that the stress-frequency coefficient of stress-compensated (SC)-cut quartz is nearly zero. Therefore, the cuts near SC-cut quartz are known to exhibit enhanced performance metrics in terms of force-frequency, resonance amplitude-frequency, acceleration-frequency, intermodulation, and dynamic thermal-frequency stability [5,29,30].

The oscillator market has recently begun to accept silicon-based MEMS resonators as an alternative for the long-stablished quartz resonators [31,32]. In theory, MEMS resonators should exhibit lower sensitivity to acceleration compared with quartz resonators because of their smaller size and mass. It is of great importance to study the physics of vibration sensitivity in silicon-based resonators in order to understand the limits and to develop proper design guidelines to reduce vibration induced instability in silicon-MEMS oscillators. Several factors such as shape, position, and number of suspension tethers, aspect ratio of the resonator dimensions, device layer thickness, and mode shape can affect the acceleration sensitivity of a MEMS resonator. Different studies have been performed to analyze the acceleration sensitivity of these resonators. In [33], the acceleration sensitivity of a lame-mode resonator supported with four anchors located at four corners of the square-shape structure is studied. In this work, it is shown that adding an extra anchor at the center of the resonator is effective at reducing the acceleration sensitivity. The acceleration sensitivity of a small-gap capacitive MEMS oscillator is studied in [34]. In capacitive resonators, the vibration-induced variation of the gap size and the transducer capacitance overlap area results in nonlinear changes in electrostatic stiffness, which results in a shift in resonance frequency.

This paper focuses on the acceleration sensitivity of silicon-based MEMS resonators. It is shown that the acceleration sensitivity of silicon-based resonators is orientation-dependent and has a correlation with the elastic properties of silicon including the Poisson ratio and nonlinear elastic coefficients. A finite element model in COMSOL is developed that predicts the hypothesized orientation-dependent acceleration sensitivity, and presented experimental results agree with the hypothesis as well. Recently, the preliminary results suggesting the orientation-dependency of the vibration sensitivity for silicon-based MEMS resonators was presented by the authors [35]. This work is an extension of the work that confirms the earlier results with a novel finite element model (FEM) and an observation of the same trends on more resonator samples.

This paper is organized as follows. The theory of acceleration sensitivity is presented in Section 2. In Section 3, the resonators used in this work are introduced and characterized. The finite element model for simulating acceleration sensitivity is explained in Section 4. Experimental results for both nonlinearity and vibration sensitivity measurements are demonstrated in Section 5. Finally, the study is summed up in Section 6.

2. Theory of Acceleration Sensitivity

Acceleration sensitivity can be defined as frequency instability of a resonator due to the mechanical vibration introduced by the environment. Applied acceleration will act on the resonator mass to produce a force, and it consequently induces stress and strain within the resonant body. There are two main causes of vibration sensitivity: geometric nonlinearity and material nonlinearity. If the applied strain due to the acceleration is large enough, the device dimensions/geometry will change (i.e., geometric nonlinearity) and consequently the resonance frequency will be affected. In addition, if the elastic stiffness coefficients of the material are strain-dependent (i.e., material elastic nonlinearity), the effective stiffness of the resonator changes due to the applied vibration and the natural resonance frequency will change as a result.

The change in the natural resonance frequency of device as a function of the applied acceleration can be represented as follows [8]:

$$f\left(\vec{a}\right) = f_0\left(1 + \vec{\Gamma}\cdot\vec{a}\right) \tag{1}$$

in which f_0 is the natural resonance frequency when there is no acceleration, $\vec{\Gamma}$ is the acceleration sensitivity vector, and $\vec{a}$ is the applied acceleration in a certain direction.

Using Equation (1), $\vec{\Gamma}$ can be found by

$$\vec{\Gamma}\cdot\vec{a} = \Delta f / f_0 \tag{2}$$

in which Δf is the shift in resonance frequency due to the applied vibration. If the applied acceleration is a constant DC vibration, a constant shift will happen at resonance frequency. However, a sinusoidal acceleration will result in a frequency modulation, and the resonance frequency will change periodically from $f_0 - \Delta f$ to $f_0 + \Delta f$ at a rate equal to the vibration frequency. Since Δf is normally a very small value, it is not easily detectable. Hence, to measure acceleration sensitivity, the resonator should be employed in an oscillator circuit. Once the output power spectrum of the oscillator is analyzed, the sidebands that appear at an offset equal to the vibration frequency of the carrier power are detectable. Let us assume the output voltage of the oscillator is given by

$$V(t) = V_0 \cos(\phi(t)) \tag{3}$$

in which $\phi(t)$ is the phase of the oscillator, which can be found by integrating the frequency

$$\phi(t) = 2\pi \int_{t_0}^{t} f(t')dt' \tag{4}$$

in which *f(t)* is the frequency of the oscillator under applied vibration. Considering a sinusoidal acceleration Equation (5), resonance frequency will be modified as shown in Equation (6)

$$\vec{a} = \vec{A}\cos(2\pi f_v t) \tag{5}$$

$$f\left(\vec{a}\right) = f_0\left(1 + \vec{\Gamma}\cdot\vec{A}\cos(2\pi f_v t)\right) \tag{6}$$

Therefore, the phase is calculated as

$$\phi(t) = 2\pi f_0 t + \left(\frac{\Delta f}{f_v}\right)\sin(2\pi f_v t), \Delta f = f_0(\vec{\Gamma}\cdot\vec{A}) \tag{7}$$

By substituting Equation (7) in Equation (3) the output voltage of the oscillator is calculated as follows:

$$V(t) = V_0\cos\left(2\pi f_0 t + \left(\frac{\Delta f}{f_v}\right)\sin(2\pi f_v t)\right) \tag{8}$$

As seen in Equation (8), the output of the oscillator under a sinusoidal vibration is a frequency-modulated signal that can be expanded using a series of Bessel functions as follows:

$$V(t) = V_0[J_0(\beta)\cos(2\pi f_0 t) + J_1(\beta)\cos(2\pi(f_0 + f_v)t) + J_1(\beta)\cos(2\pi(f_0 - f_v)t) + \ldots] \tag{9}$$

in which $\beta = \Delta f / f_v = (\vec{\Gamma}\cdot\vec{A})\,f_0/f_v$ is the modulation index. For $\beta < 0.1$, $J_0(\beta) = 1$, $J_1(\beta) = \beta/2$, and $J_n(\beta) = 0$ for $n > 2$. Hence, for small modulation indices, the output spectrum of the oscillator only contains the main resonance frequency and two sidelobes at $f_0 \pm f_v$. As seen from Equation (9), the magnitude of these sidelobes ($\beta/2$) depends on the acceleration amplitude and frequency, resonance frequency of the resonator, and the acceleration sensitivity vector. The ratio of the power in these two sidelobes to the power of the carrier is derived through the following equation

$$L_v = \left(\frac{J_1(\beta)}{J_0(\beta)}\right)^2 \tag{10}$$

or equivalently in decibels notation

$$L_{v,dB} = 20\log\left(\frac{J_1(\beta)}{J_0(\beta)}\right) = 20\log\left(\frac{\beta}{2}\right) = 20\log\left(\frac{\left(\vec{\Gamma}\cdot\vec{A}\right)f_0}{2f_v}\right) \tag{11}$$

So, acceleration sensitivity in *i*th direction, $\vec{\Gamma}$, can be found as follows:

$$\Gamma_i = (2f_v/A_i f_0)10^{L_v/20} \tag{12}$$

in which f_v is the vibration frequency, A_i is the amplitude of vibration in *i*th direction, and L_v is the difference between the carrier signal power and sidebands power in *dB*.

The amplitude of total acceleration sensitivity along x, y, and z direction is calculated as follows:

$$\Gamma_t = \sqrt{\Gamma_x^2 + \Gamma_y^2 + \Gamma_z^2} \tag{13}$$

3. Resonator Design and Characterization

Thin-film piezoelectric-on-substrate (TPoS) platform is chosen in this work for prototyping the silicon-based resonators as they offer high quality factor and low insertion loss [36]. The schematic of a typical TPoS resonator used in this work is shown in Figure 1, in which a thin sputtered piezoelectric AlN layer, sandwiched between top and bottom electrodes made of Molybdenum, is stacked on top of a <100> silicon layer. The top electrode is patterned on the resonant structure to enable two-port operation of the resonator. The specific interdigitated configuration could be used to excite higher harmonic lateral-extensional resonance modes [37,38]. However, in this work the first harmonic resonance is used in the oscillator, and the top electrodes are connected to each other for the resonator to be operated in a one-port configuration. In order to study the effect of orientation on acceleration sensitivity, resonators with identical design are rotated to align to different crystalline orientations. The stack layers of the resonators used in this work include 8 μm silicon, 1 μm AlN, and 100 nm Molybdenum used as electrodes. The substrate for all resonators is a phosphorus-doped silicon-on-insulator (SOI) wafer with a device layer doped at a concentration of ~5 × 10^{19} cm^{-3}.

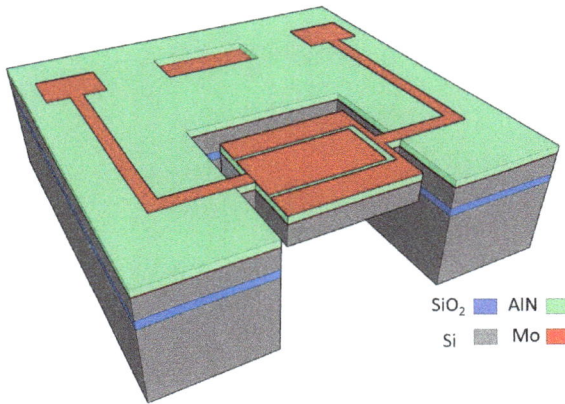

Figure 1. The schematic of a typical TPoS resonator used in this work.

A schematic representing the relative position of the resonators aligned to <100> and <110> orientations on the [100] SOI wafer is shown in Figure 2. The fundamental lateral extensional mode-shape of the resonant structure used in this work is also shown in this figure.

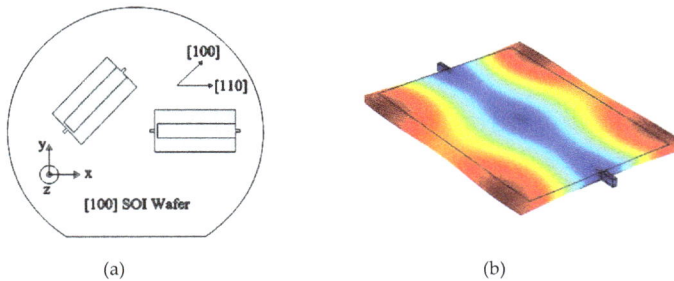

(a)   (b)

Figure 2. (**a**) A schematic representation of the relative position of <100> and <110> resonators on the [100] wafer and (**b**) the simulated lateral extensional mode-shape of a resonator used in this work.

The frequency response (S21 amplitude) and the scanning electron micrograph (SEM) of the resonators used in this study are also shown in Figure 3.

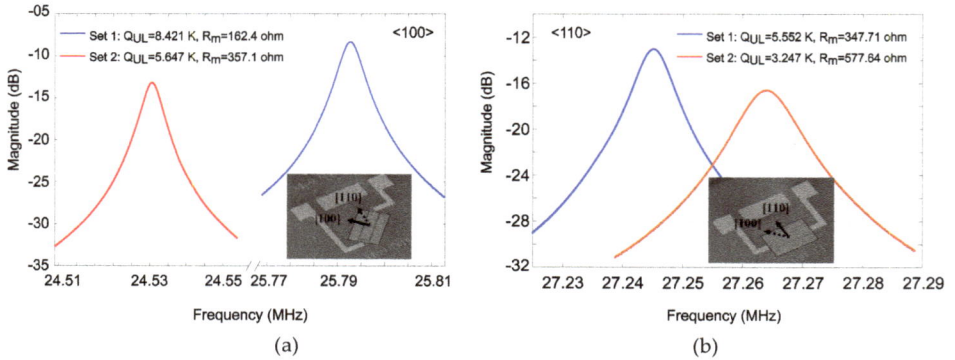

Figure 3. The frequency response (i.e., S21) and the SEM of the two sets of resonators used in this work aligned to <100> (**a**) and <110> (**b**) silicon planes.

4. Finite Element Simulation

A theoretical or simulation model that could be used to predict and improve acceleration sensitivity of resonators is very desirable for design purposes. Due to the nonlinear nature of vibration sensitivity, developing such tools is not straightforward. Complicated equations have been developed for vibration sensitivity of bulk mode resonators, indicating dependency of vibration sensitivity to a complex function of linear and nonlinear elastic constants [39]. However, a finite element model can be more useful, as numerical solutions have recently found popularity due to the great progress made in computational speed and capacity.

In order to develop a finite element model for this purpose, geometric and material nonlinearities should be both included. Geometric nonlinearity is already predefined in commercial FEM software packages such as COMSOL, but capturing material nonlinearity of silicon is more challenging.

4.1. Geometric Nonlinearity

When there is relatively large deformation and rotation, the approximations typically used in linear elastic equations break apart, and it is necessary to identify the distinction between deformed and undeformed configuration, and the nonlinear definitions of stress and strain should be utilized as opposed to the simplified engineering definitions.

The definition of linear engineering stain is shown in Equation (14), which needs to be replaced by Equation (15), the nonlinear Green–Lagrangian strain, in case of having rotation or large deformation.

$$e_X = \frac{\partial u}{\partial X} \tag{14}$$

$$S_X = \frac{\partial u}{\partial X} + \frac{1}{2}\left(\frac{\partial u}{\partial X}\right)^2 + \frac{1}{2}\left(\frac{\partial V}{\partial X}\right)^2 + \frac{1}{2}\left(\frac{\partial w}{\partial X}\right)^2 \tag{15}$$

Therefore, it is required to modify the equations of the FEM simulator in order to consider nonlinear definitions of strain. COMSOL software will provide the option to consider geometric nonlinearity in the model, and, once selected, the Green–Lagrangian strain will be utilized in all calculations.

4.2. Material Nonlinearity

In general, there is a nonlinear relationship between stress and strain for any arbitrary material. Normally, for small strain values, one could use a linear approximate instead of the nonlinear relation. However, for large values of strain, the nonlinear equation should be considered. In the case of simulating acceleration sensitivity, we are dealing with small values of resonance frequency shift, in the range of ppb. Thus, considering nonlinear terms is critical in accurately predicting very small changes in resonance frequency, even though the applied acceleration and the resulted strain are small. Equation (16) shows the nonlinear relationship between second Piola–Kirchhoff (2nd PK) stress and Green–Lagrangian strain for silicon [40,41]

$$T_{ij}(X) = C_{ijkl}S_{kl} + \frac{1}{2}C_{ijklmn}S_{kl}S_{mn} \tag{16}$$

in which C_{ijkl}, C_{ijklmn}, T_{ij}, and S_{kl} are linear and nonlinear elastic stiffness coefficients, 2nd PK stress, and Green–Lagrangian strain, respectively. In order to capture acceleration sensitivity of the resonators correctly, we propose to define this nonlinear stress-strain relation in COMSOL as follows. First, Equation (16) will be rewritten as

$$T_{ij}(X) = (C_{ijkl} + \frac{1}{2}C_{ijklmn}S_{mn})S_{kl} \tag{17}$$

Next, the expression in parenthesis could be defined as an elastic stiffness tensor:

$$C_{ijkl}' = C_{ijkl} + \frac{1}{2}C_{ijklmn}S_{mn}, \; i,j,k,l = 1, 2, 3 \tag{18}$$

With this new definition, the linear second-order elastic stiffness tensor can be modified so that it is strain-dependent using the coefficients of nonlinear elasticity. Since working with tensors is not straight forward, Voigt notation should be used to convert tensorial format to matrix format. The same concept presented above is shown below in matrix format:

$$T_i(X) = (C_{ij} + \frac{1}{2}C_{ijk}S_k)S_j \tag{19}$$

$$C_{ij}' = C_{ij} + \frac{1}{2}C_{ijk}S_k \qquad i,j,k = 1, 2, 3, 4, 5, 6$$

Now using Einstein notation and considering cubic symmetry for silicon structure, the equation above can be expanded to find all of the 36 components of the new modified stiffness matrix C_{ij}'.

The structure used in this work for modeling is shown in Figure 4, which includes the lateral extensional resonator connected to the substrate. It should be noted that in the proposed FEM model, only the silicon layer is considered. Utilizing such simplified model is justified as, firstly, the thickness of the AlN layer is 1/8th of the thickness of the silicon layer, and, secondly, the sputtered AlN used in these resonators is isotropic in x-y plane. Therefore, it is concluded that the orientation dependency of acceleration sensitivity in resonators is mainly dominated by their structural silicon body.

To study vibration sensitivity using COMSOL, a prestressed analysis is performed on the discussed structure as follows: first, a stationary study is performed to obtain the final strain values due to the applied acceleration followed by a second step of eigenfrequency study to predict the modified natural resonance frequency while the initial strains calculated in the first step are applied to the structure. It should be noted that acceleration is applied as a body load equal to $\vec{F} = \rho\vec{a}$ (N/m^3) to the whole structure including the frame, in which ρ is effective density. It is worth mentioning that COMSOL uses a perturbation model to calculate the eigenfrequencies of the pre-stressed system. Although it is possible to directly solve a set of differential equations to find the resonance frequency under acceleration but as described in [39], it has been shown that calculating the first eigenvalue perturbation is considerably more efficient.

Acceleration sensitivity is calculated as normalized resonance frequency shift due to the applied acceleration using Equation (2). The resulted vibration sensitivity in x, y, and z direction as a function of acceleration amplitude is shown in Figure 5 for a resonator aligned to <100> orientation. As seen, the acceleration sensitivity in z direction is the dominant component of the Γ vector, which is expected, as in this case vibration is applied normally to the resonator plane, introducing an out-of-plane bending moment with a relatively low effective stiffness in the structure. Hence, only acceleration sensitivity in z direction is demonstrated in the following figures for simplicity. Material nonlinearity for <100> orientation is defined in COMSOL, as explained before. In order to consider nonlinearity in <110> orientation, a 45° rotated coordinate system in the resonator plane is used. The simulated acceleration sensitivity for both resonators aligned to <100> and <110> orientations considering linear and nonlinear material properties is demonstrated in Figure 6.

Figure 4. The structure developed in COMSOL for modeling the acceleration sensitivity.

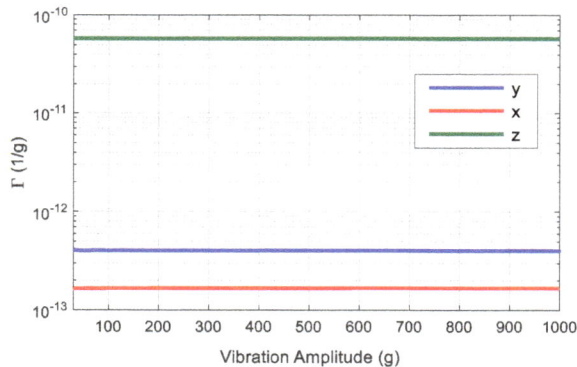

Figure 5. The simulated acceleration sensitivity in x, y, and z directions for a resonator with lateral extensional mode-shape aligned to <100> orientation.

As seen, acceleration sensitivity of the <100> resonator is larger than the <110> resonator, even for the case of linear material properties. One possible reason is that the Young's modulus in <100> silicon is smaller and the Poisson ratio is larger than that of the <110> orientation [42]. Hence, when a load is applied in <100> direction normal to the plane of the resonator, larger bending results, and the in-plane deformation of the resonator would be larger for the resonator aligned to <100> plane compared with the resonator aligned to <110> plane. The resulting in-plane strain will directly affect the resonance frequency, as the frequency is a function of the resonator's in-plane dimensions. Therefore, it appears that the dominant factor in the acceleration sensitivity of the resonator studied in this work is the Poisson ratio of silicon. However, it should be noted that material nonlinearity also affects the absolute value of acceleration sensitivity significantly. As seen, when material nonlinearity is considered, acceleration sensitivity is increased in both <100> and <110> cases. When silicon is defined as a linear material, the ratio of vibration sensitivity of resonator aligned with <100> to that aligned with <110> is 4.7, while this ratio increases to 18.5 when the material nonlinearity of silicon is considered. Therefore, material nonlinearity significantly affects both the value and the orientation dependency of the acceleration sensitivity.

It is worth mentioning that this simulation only considers the resonator and part of the substrate that is simplified compared to the experimental setups. In addition, elastic constants used in this simulation are for n-type silicon with doping concentration of 2×10^{19} cm^{-3} [43], which is the closest available data to the doping level for the devices fabricated in this work ($\sim 5 \times 10^{19}$ cm^{-3}). Furthermore, defining material nonlinearity for other layers of the device including AlN may affect the absolute value of acceleration sensitivity. However, in this study, the goal is to investigate the effect of orientation on acceleration sensitivity of n-type highly doped silicon-based resonators and understand the effect of material nonlinearity on that. Thus, predicting the relative sensitivity in different orientations is the most important aim of this simulation, which can be achieved with this simplified model.

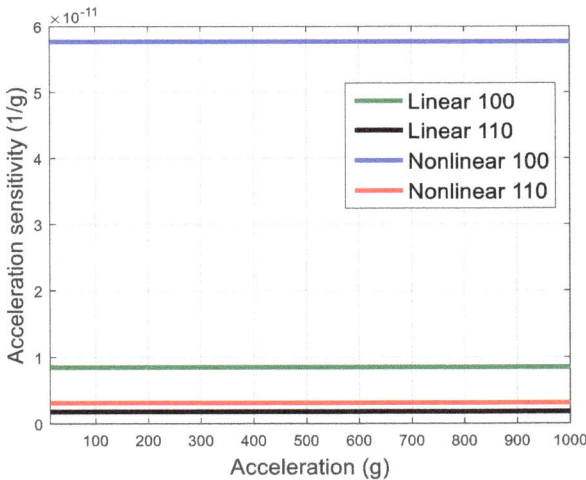

Figure 6. The simulated acceleration sensitivity considering linear/nonlinear elastic properties for both <100> and <110> crystalline orientations.

From the above simulation, it is concluded that by increasing material nonlinearity, acceleration sensitivity will increase. To confirm this conclusion, nonlinear elastic coefficients of <100> silicon are intentionally modified so that a larger amplitude frequency, *A-f*, coefficient is obtained. This coefficient, *k*, is a measure of nonlinearity, which will be explained in more detail in Section 5. In order to calculate *k* as a function of nonlinear stiffness constants, first, the nonlinear Young's modulus is calculated based on the closed form equations presented in [40]. Then, amplitude-frequency coefficient *k* is calculated using equations presented in [44] for lateral extensional mode-shape aligned with <100> orientation.

The simulated acceleration sensitivity using the modified elastic constants is shown in Figure 7. As seen, by increasing k (i.e., nonlinearity), acceleration sensitivity also increases, which confirms the correlation between acceleration sensitivity and nonlinearity.

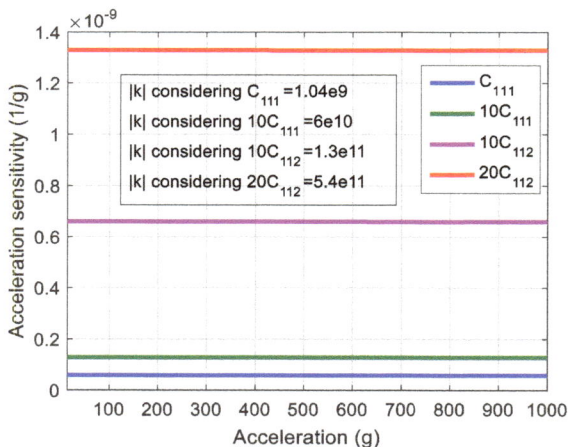

Figure 7. The acceleration sensitivity of <100> resonator using modified elastic constants, indicating the correlation between elastic nonlinearity and acceleration sensitivity.

5. Measurements and Results

As explained in Section 1, we hypothesized that acceleration sensitivity of silicon-based MEMS resonators is orientation-dependent and affected by the elastic properties of the resonator material. In order to study this hypothesize, both nonlinearity and acceleration sensitivity of above resonators need to be measured.

5.1. Nonlinearity Measurements

Amplitude-frequency (*A-f*) coefficient k is a measure of nonlinearity in resonators. This coefficient determines the relation between normalized resonance frequency shift of the resonator and the amplitude of vibration (x) as presented by (20).

$$f = f_0\left(1 + kx^2\right) \tag{20}$$

We will use backbone curve plots obtained through the ringdown response measurement in order to evaluate nonlinearity of the resonators. A backbone curve is the plot of normalized frequency shift versus amplitude of vibration. In order to find the shift of resonance frequency in nonlinear regime, two methods can be used: forced or unforced oscillation. In former, the input excitation power is increased gradually, and the shift in resonance frequency is read using a network analyzer while the resonator is under forced excitation. However, in the latter, ringdown response of the resonator is measured, and by analyzing the decaying signal, shift in natural resonance frequency of the unforced device vibration can be calculated [45]. In this study, second method is used.

Each resonator is forced by a large enough input that guarantees a detectable shift in resonance frequency. After exciting the resonator for a period of time, the input power is ceased, and the unforced ringdown response of the device is captured. The time domain behavior of the output voltage signal for one of the resonators used in this study captured by a digital oscilloscope is demonstrated in Figure 8.

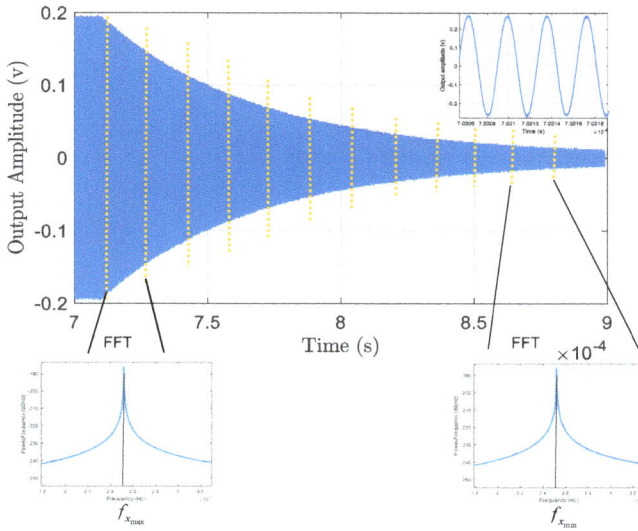

Figure 8. The ringdown response of the resonator and the FFT plots for different sections of the decaying signal to find the backbone curve. $f_{x_{max}}$ and $f_{x_{min}}$ are the resonance frequencies of the device corresponding to largest and smallest amplitudes of vibration, respectively.

The decaying signal is then spilt into several bins, and a fast Fourier transform (FFT) is used for the data in each bin to find the natural resonance frequency of the resonator corresponding to the varying output voltage (vibration amplitude) during the ringdown signal (Figure 8). Equivalently, instead of using FFT over one decaying signal, the resonator can be excited separately with different input powers. Then, for each input power FFT is taken over the first bin of decaying signal. Figure 9 shows the ringdown response for different input powers.

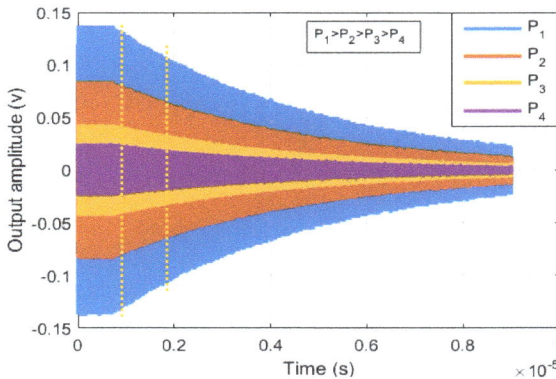

Figure 9. The ringdown response of the resonator for different input powers. The resonance frequencies corresponding to the highest and lowest power (P_1 and P_4) are $f_{x_{max}}$ and $f_{x_{min}}$, respectively.

Now, in order to plot the backbone curve, the output voltage needs to be converted to the amplitude of vibration. For capacitive resonators, there already are closed form equations developed to perform this conversion [45]. However, for piezoelectric resonators such equations are not available in literature. So to find the amplitude of vibration for each input power, the S parameters of the resonators are collected for each case. Then, energy stored in resonator in each case is calculated using the method proposed in [46]. In this approach, the magnitude and phase of the input and output current and voltage of the resonator are calculated using a model developed in the advanced design system (ADS) software and the S-parameters of the device. The energy stored in resonator is then obtained using those calculated parameters [46]. This energy can be approximated by

$$E = \frac{1}{2}k_0 x^2 \tag{21}$$

in which x is the amplitude of vibration and k_0 is the linear spring stiffness constant of the resonator, which for lateral-extensional mode-shape can be calculated by [44]

$$k_0 = \frac{\pi^2 E A}{2L} \tag{22}$$

in which E is the effective Young's modulus, and A and L are the cross section area and length of the resonator, respectively. Using Equation (21), amplitude of vibration for each input power is calculated, and the backbone curves are plotted. The plot is then fitted with Equation (23) in order to find *A-f* coefficient k.

$$\Delta f / f_0 = k x^2 \tag{23}$$

The backbone curve and the fitted plot for both <100> and <110> resonators in set 1 is shown in Figure 10. As seen, the absolute value of k for <100> resonator (4.803) is more than 3× larger than that calculated for the <110> resonator (1.411). This confirms a larger nonlinearity for the <100> resonator.

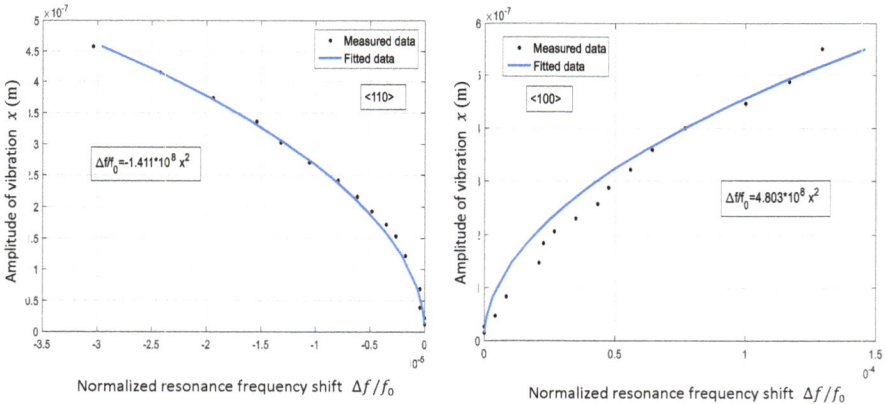

Figure 10. The backbone curve for <110> (**left**) and <100> (**right**) resonators and the associated amplitude-frequency (k) coefficients calculated based on the curves, indicating higher nonlinearity in the <100> resonator.

Hence, this resonator is expected to be more sensitive to acceleration, as simulation predicts more sensitivity by increasing nonlinearity.

5.2. Acceleration Sensitivity Measurements

In order to measure acceleration sensitivity, the resonator must be employed in an oscillator circuit. To minimize the number of components, a commercial oscillator IC, CF5027 is used. The printed circuit board (PCB) and the schematic of the oscillator used in this study are shown in Figure 11a,b. The connections XT and XTN denoted on the IC are the amplifier input and output, into which the MEMS resonator should be connected. Details of the oscillator IC specifications can be found in the data sheet [47].

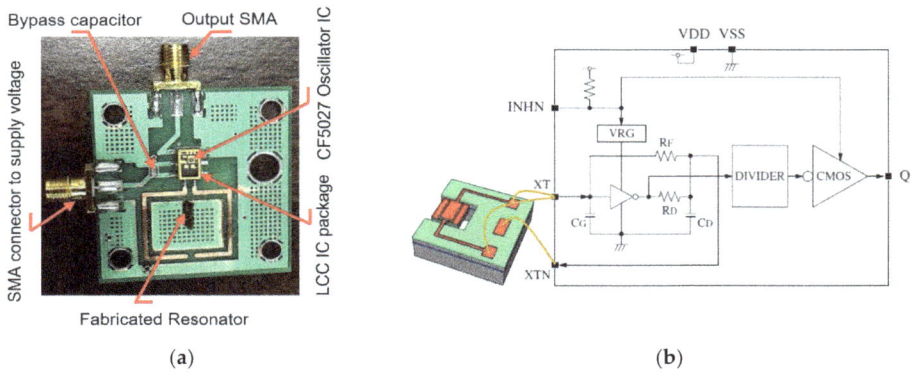

(a)   (b)

Figure 11. (**a**) The oscillator PCB containing the commercial oscillator IC (CF5027) and (**b**) the schematic of the oscillator IC connected to the TPoS resonator [47].

The typical phase noise measured from the assembled oscillators is shown in Figure 12. As seen, both oscillators exhibit excellent phase noise performance, which enables accurate measurement of the acceleration sensitivity.

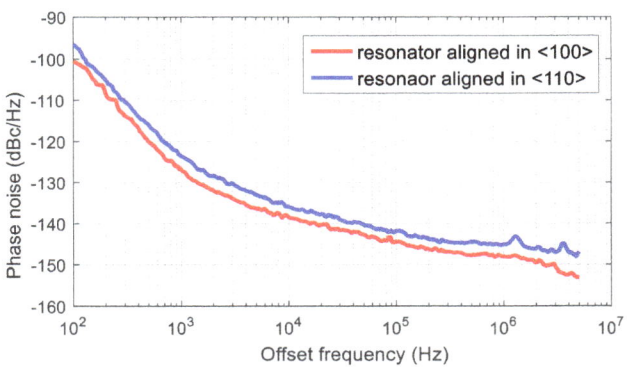

Figure 12. The phase noise performance of the oscillator circuits containing the <100> and <110> resonators.

A closed loop magnetic shaker system from Vibration Research Group Inc. is used to simulate the actual vibration coming from environment (Figure 13). The acceleration applied to the board is measured by a DYTRAN 3055D1T accelerometer. The sensed acceleration is then fed back to the controller to be compared with the desired acceleration set by the operator. A control signal is afterward sent to the amplifier, which determines the power generated by the amplifier that feeds the magnetic motor.

Figure 13. The acceleration-sensitivity measurement setup.

The phase noise and the output spectrum of the two resonators aligned with <100> and <110> orientations are shown in Figure 14 when a 14 g, 800 Hz sinusoidal acceleration is applied in z direction. As seen, two sidebands appear at an offset frequency of 800 Hz from the carrier in the oscillator output spectrum for both resonators, as expected from theoretical explanations provided in Section 2. The amplitude of these sidebands is proportional to the acceleration sensitivity of resonator. Hence, it is obvious that resonator aligned to <100> plane is significantly more sensitive to the acceleration compared with the one aligned to the <110> plane. Acceleration sensitivity of these devices is 3.1×10^{-9} and 4.8×10^{-8} 1/g for resonators aligned with <110> and <100> orientations, respectively. Spurious signals will also appear on the phase noise plot at an offset frequency equal to the vibration frequency. In fact, the sidebands on the output spectrum are considered as noise, and hence the corresponding spurs on the phase noise are expected. It should be noted that the amplitude of the spurs on the phase noise is almost equal to the difference between amplitude of the carrier and sidebands in the spectrum, which corresponds to how the phase noise is calculated.

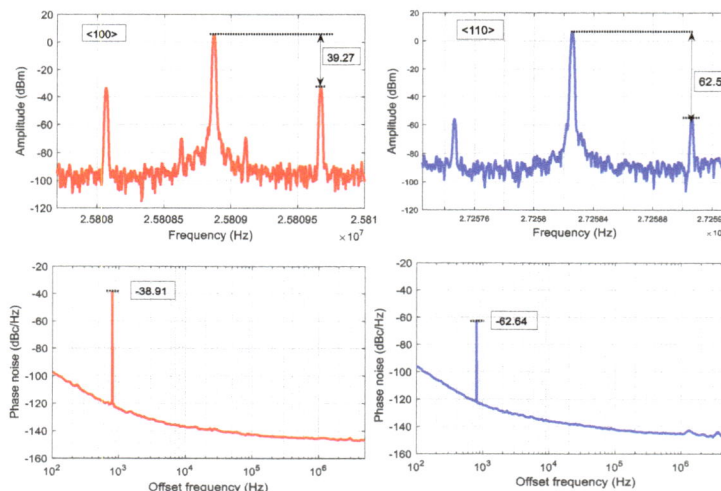

Figure 14. The oscillator output spectrum and the phase noise under 14 g, 800 Hz vibration for set 1 of resonators. Larger amplitude of the sidebands that appeared at 800 Hz offset frequency from the carrier in the output frequency spectrum of the oscillator with <100> resonator shows higher acceleration sensitivity of this device. More theoretical details are provided in Section 2.

It is worth mentioning that, in this study, the phase noise and frequency spectrum measurements are done with a Rohde & Schwarz signal source analyzer. However, one can configure a platform to do the same measurement, such as those presented in [48,49].

Acceleration sensitivity of all resonators has been measured by sweeping both acceleration amplitude and frequency. The results for first set of resonators are shown in Figure 15. By averaging total acceleration sensitivity over vibration frequency range of 0–3 kHz, ~5.66 × 10^{-8} and ~3.66 × 10^{-9} are obtained for resonators aligned to <100> and <110> orientations, respectively.

It is worth mentioning that, contrary to the measurement, the acceleration sensitivity is not expected to vary with acceleration frequency. This is because the induced strain in the resonant structure should not be a function of the acceleration frequency, as no resonance modes of the device are within the range of applied frequency. However, the acceleration is not directly applied to the device, and, rather, it is being applied to a board to which the resonator die is attached. It is suspected that there are numerous resonance modes for the entirety of the board and the components it contains including the wirebonds, cables, and connectors. Therefore, the effective force applied to the resonator will end up varying with the frequency. In addition, other sources of acceleration sensitivity such as a shift in the phase can play a role that is frequency-dependent.

The same measurement has been repeated for the second set of devices, and the same trend as for the first is obtained, confirming our hypothesis on orientation-dependency of acceleration sensitivity. Acceleration sensitivity of both sets of resonators for a 7 g, 2800 Hz sinusoidal vibration is reported in Table 1.

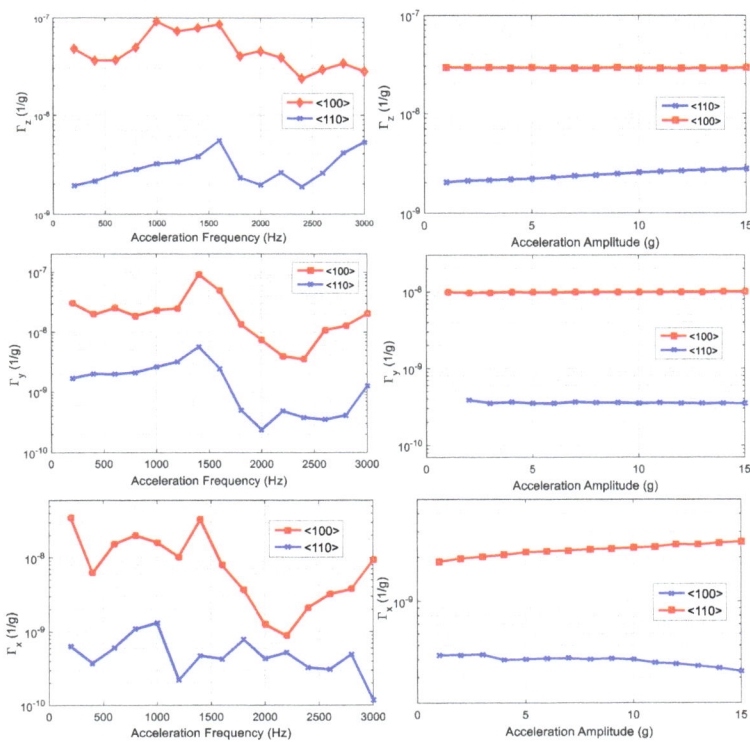

Figure 15. Acceleration sensitivity in x, y, z direction vs. acceleration frequency (**left**) and magnitude (**right**) for set 1 of resonators.

Table 1. Acceleration sensitivity for both sets of devices for a 7 g, 2800 Hz sinusoidal vibration.

	Orientation	Γ_x	Γ_y	Γ_z	Γ_{total}
Set 1	<100>	3.73×10^{-9}	1.2×10^{-8}	3.3×10^{-8}	3.6×10^{-8}
	<110>	4.9×10^{-10}	4.1×10^{-10}	4×10^{-9}	4.1×10^{-9}
Set 2	<100>	8.5×10^{-9}	2.5×10^{-8}	8.1×10^{-8}	8.5×10^{-8}
	<110>	3.1×10^{-10}	6.4×10^{-9}	4.8×10^{-9}	8×10^{-9}

6. Conclusions

The acceleration sensitivity of n-type, highly-doped, silicon-based extensional resonators aligned with different crystalline orientations is studied. Experimental results suggest that a resonator aligned with the <110> plane direction is much less sensitive to applied acceleration compared with a similar resonator aligned with <100> plane. A finite element model is presented to simulate the acceleration sensitivity of the resonators used in this study. Simulation results also confirm less sensitivity for the <110> resonator. In addition, it was shown that material nonlinearity is an important factor affecting the acceleration sensitivity of these type of resonators.

Author Contributions: B.K. developed the idea, generated the simulation results, collected the experimental data, and wrote the paper; J.G. fabricated the resonators; R.A. contributed in developing the idea and revised the manuscript; R.A. and B.K. analyzed the results.

Conflicts of Interest: The authors declare no conflicts of interest.

References

1. Ng, E.; Yang, Y.; Hong, V.A.; Ahn, C.H.; Heinz, D.B.; Flader, I.; Chen, Y.; Everhart, C.L.M.; Kim, B.; Melamud, R.; et al. The long path from MEMS resonators to timing products. In Proceedings of the 28th IEEE International Conference on Micro Electro Mechanical Systems (MEMS), Estoril, Portugal, 18–22 January 2015; pp. 1–2.

2. Abdolvand, R.; Bahreyni, B.; Lee, J.E.; Nabki, F. Micromachined resonators: A review. *Micromachines* **2016**, *7*. [CrossRef]

3. Beek, J.T.M.V.; Puers, R. A review of MEMS oscillators for frequency reference and timing applications. *J. Micromech. Microeng.* **2011**, *22*, 013001. [CrossRef]

4. Walls, F.L.; Gagnepain, J.-J. Environmental sensitivities of quartz oscillators. *IEEE Trans. Ultrasonics Ferroelectr. Freq. Control* **1992**, *392*, 241–249. [CrossRef] [PubMed]

5. Ballato, A.; Mizan, M. Simplified expressions for the stress-frequency coefficients of quartz plates. *IEEE Trans. Ultrasonics Ferroelectr. Freq. Control* **1984**, *311*, 11–17. [CrossRef]

6. Anderson, C.L.; Bagby, J.S.; Barrett, R.L.; Ungvichian, V. Acceleration charge sensitivity in AT-quartz resonators. In Proceedings of the IEEE International Frequency Control Symposium, San Francisco, CA, USA, 31 May–2 June 1995.

7. Yong, Y.K.; Chen, J. Effects of initial nonlinear strains and nonlinear elastic constants in force-frequency and acceleration sensitivity of quartz resonators. In Proceedings of the IEEE International Frequency Control Symposium (IFCS), New Orleans, LA, USA, 9–12 May 2016; pp. 1–2.

8. Filler, R.L. The acceleration sensitivity of quartz crystal oscillators: A review. *IEEE Trans. Ultrasonics Ferroelectr. Freq. Control* **1988**, *35*, 297–305. [CrossRef] [PubMed]

9. Chen, J.; Yong, Y.K.; Kubena, R.; Kirby, D.; Chang, D. On the acceleration sensitivity and its active reduction by edge electrodes in at-cut Quartz resonators. *IEEE Trans. Ultrasonics Ferroelectr. Freq. Control* **2015**, *626*, 1104–1113. [CrossRef] [PubMed]

10. Chen, J.; Yong, Y.K.; Kubena, R.; Kirby, D.; Chang, D. Nonlinear acceleration sensitivity of quartz resonators. In Proceedings of the Joint Conference of the IEEE International Frequency Control Symposium & the European Frequency and Time Forum, Denver, CO, USA, 12–16 April 2015; pp. 11–16.

11. Fry, S.J.; Burnett, G.A. Reducing the acceleration sensitivity of AT-strip quartz crystal oscillators. In Proceedings of the International Frequency Control Symposium, Newport Beach, CA, USA, 1–4 June 2010; pp. 25–30.

12. Rosati, V.R.; Filler, R.L. Reduction of the effects of vibration on SC-cut quartz crystal oscillators. In Proceedings of the IEEE 35th Annual Frequency Control Symposium, Philadelphia, PA, USA, 27–29 May 1981.

13. Fruehauf, H. *"g"- Compensated, Miniature, High Performance Quartz Crystal Oscillators*; Frequency Electronics Inc.: New York, NY, USA, 2007.

14. Bloch, M.; Mancini, O.; McClelland, T.; Terracciano, L. *Acceleration "G" Compensated Quartz Crystal Oscillators*; Frequency Electronics, Inc.: New York, NY, USA, 2009.

15. Bhaskar, S.; Curran, J.T.; Lachapelle, G. Improving GNSS carrier-phase tracking via oscillator g-sensitivity compensation. *IEEE Trans. Aerosp. Electron. Syst.* **2015**, *514*, 2641–2654. [CrossRef]

16. Hati, A.; Nelson, C.W.; Taylor, J.; Ashby, N.; Howe, D.A. Cancellation of Vibration Induced Phase Noise in Optical Fibers. *IEEE Photonics Technol. Lett.* **2008**, *20*, 1842–1844. [CrossRef]

17. Walls, F.L.; Vig, J.R. Acceleration Insensitive Oscillator. U.S. Patent 4.575.690, 1985.

18. Fry, S.; Bolton, W.; Esterline, J. Crystal Oscillator with Reduced Acceleration Sensitivity. U.S. Grant US8525607B2, 2008.

19. Ballato, A. Resonators compensated for acceleration fields. In Proceedings of the IEEE 33rd Annual Symposium on Frequency Control, Atlantic City, NJ, USA, 30 May–1 June 1979.

20. Besson, R.J.; Peier, U.R. Further advances on BVA quartz resonators. In Proceedings of the IEEE 34th Annual Symposium on Frequency Control, Philadelphia, PA, USA, 28–30 May 1980.

21. Hati, A.; Nelson, C.; Howe, D. *Vibration-induced PM Noise in Oscillators and its Suppression*; National Institute of Standards and Technology: Gaithersburg, MD, USA, 2009.

22. Maleki, L. High Performance Optical Oscillators for Microwave and mm-wave Applications. *Microw. J.* **2013**, *56*, 106.

23. Leibrandt, D.R.; Bergquist, J.C.; Rosenband, T. Cavity-stabilized laser with acceleration sensitivity below $10^{-12}/\mathrm{g}^{-1}$. *Phys. Rev. A* **2013**, *87*, 023829. [CrossRef]

24. Ludlow, A.D.; Hung, X.; Notcutt, M.; Zanon-Willette, T.; Foreman, S.M.; Boyd, M.M.; Blatt, S.; Ye, J. Compact, thermal-noise-limited optical cavity for diode laser stabilization at 1×10^{-15}. *Opt. Lett.* **2007**, *32*, 641–643. [CrossRef] [PubMed]

25. Millo, J.; Dawkins, S.; Chicireanu, R.; Magalhaes, D.V.; Mandache, C.; Holleville, D.; Lours, M.; Bize, S.; Lemonde, P.; Santarelli, G. Ultra-stable optical cavities: Design and experiments at LNE-SYRTE. In Proceedings of the IEEE International Frequency Control Symposium, Honolulu, HI, USA, 19–21 May 2008; pp. 110–114.

26. Webster, S.A.; Oxborrow, M.; Gill, P. Vibration insensitive optical cavity. *Phys. Rev. A* **2007**, *75*. [CrossRef]

27. Kosinski, J.A.; Ballato, A. Designing for low acceleration sensitivity. *IEEE Trans. Ultrasonics Ferroelectr. Freq. Control* **1993**, *405*, 532–537. [CrossRef] [PubMed]

28. Kosinski, J.A.; Pastore, R.A. Theory and design of piezoelectric resonators immune to acceleration: Present state of the art. *IEEE Trans. Ultrasonics Ferroelectr. Freq. Control* **2001**, *485*, 1426–1437. [CrossRef]

29. Ballato, A.; Eernisse, E.P.; Lukaszek, T.J. *Force-Frequency Effect in Doubly Rotated Quartz Resonators*; Sandia Labs: Albuquerque, NM, USA, 1977.

30. Eernisse, E.P.; Lukaszek, T.J.; Ballato, A. Variational calculation of force-frequency constants of doubly rotated quartz resonators. *IEEE Trans. Sonics Ultrasonics* **1978**, *253*, 132–137. [CrossRef]

31. SiTime corporation. MEMS Oscillators: Enabling Smaller, Lower Power IoT & Wearables. Available online: https://www.sitime.com/api/gated/SiTime-MEMS-Enable-Small-Low-Power-IoT-Wearables.pdf (accessed on 11 May 2018).

32. SiTime corporation. Increase Automotive Reliability and Performance with High-temperature, Ultra Robust MEMS Oscillators. Available online: https://www.sitime.com/api/gated/SiTime-MEMS-Oscillators-for-Automotive-Applications.pdf (accessed on 11 May 2018).

33. Kim, B.; Olsson, R.H.; Smart, K.; Wojciechowski, K.E. MEMS Resonators with Extremely Low Vibration and Shock Sensitivity. In Proceedings of the IEEE Sensors, Limerick, Ireland, 28–31 October 2011.

34. Kim, B.; Akgul, M.; Lin, Y.; Li, We.; Ren, Z.; Nguyen, C.T.-C. Acceleration sensitivity of small-gap capacitive micromechanical resonator oscillators. In Proceedings of the IEEE International Frequency Control Symposium (FCS), Newport Beach, CA, USA, 1–4 June 2010.

35. Khazaeili, B.; Abdolvand, R. Orientation-dependent acceleration sensitivity of silicon-based MEMS resonators. In Proceedings of the IEEE International Frequency Control Symposium (IFCS), New Orleans, LA, USA, 9–12 May 2016; pp. 1–5.

36. Harrington, B.P.; Shahmohammadi, M.; Abdolvand, R. Toward ultimate performance in GHz MEMS resonators: Low impedance and high Q. In Proceedings of the IEEE International Conference on Micro Electro Mechanical Systems (MEMS), Hong Kong, China, 24–28 January 2010; pp. 707–710.

37. Fatemi, H.; Shahmohammadi, M.; Abdolvand, R. Ultra-stable nonlinear thin-film piezoelectric-on-substrate oscillators operating at bifurcation. In Proceedings of the IEEE 27th International Conference on Micro Electro Mechanical Systems (MEMS), San Francisco, CA, USA, 26–30 January 2014; pp. 1285–1288.

38. Shahmohammadi, M.; Fatemi, H.; Abdolvand, R. Nonlinearity reduction in silicon resonators by doping and re-orientation. In Proceedings of the IEEE 26th International Conference on Micro Electro Mechanical Systems (MEMS), Taipei, Taiwan, 20–24 January 2013; pp. 793–796.

39. Kosinski, J.A. The fundamental nature of acceleration sensitivity. In Proceedings of the IEEE International Frequency Control Symposium, Honolulu, HI, USA, 5–7 June 1996; pp. 439–448.

40. Kim, K.Y.; Sachse, W. Nonlinear elastic equation of state of solids subjected to uniaxial homogeneous loading. *J. Mater. Sci.* **2000**, *3513*, 3197–3205. [CrossRef]

41. Kaajakari, V.; Mattila, T.; Lipsanen, A.; Oja, A. Nonlinear Mechanical Effects in Silicon Longitudinal Mode Beam Resonators. *Sens. Actuators A Phys.* **2005**, *120*, 64–70. [CrossRef]

42. Wortman, J.; Evans, R. Young's modulus, shear modulus, and Poisson's ratio in silicon and germanium. *J. Appl. Phys.* **1965**, *361*, 153–156. [CrossRef]

43. Hall, J.J. Electronic Effects in the Elastic Constants of n-Type Silicon. *Phys. Rev.* **1967**, *161*. [CrossRef]

44. Yang, Y.; Ng, E.J.; Hong, V.A.; Ahn, C.H.; Chen, Y.; Ahadi, E.; Dykman, M.; Kenny, T.W. Measurement of the nonlinear elasticity of doped bulk-mode MEMS resonators. In Proceedings of the Solid-State Sensors, Actuators and Microsystems Workshop, Hilton Head Island, SC, USA, 8–12 June 2014.

45. Yang, Y.; Ng, E.J.; Polunin, P.M.; Chen, Y.; Flader, I.B.; Shaw, S.W.; Dykman, M.I.; Kenny, T.W. Nonlinearity of Degenerately Doped Bulk-Mode Silicon MEMS Resonators. *J. Microelectromech. Syst.* **2016**, *255*, 859–869. [CrossRef]

46. Fatemi, H.; Abdolvand, R. Fracture Limit in Thin-Film Piezoelectric-on-Substrate Resonators: Silicon vs. Diamond. In Proceedings of the IEEE International Conference on Micro Electro Mechanical Systems (MEMS), Taipei, Taiwan, 20–24 January 2013; pp. 461–464.

47. *Crystal Oscillator Module ICs*; MSDS No. CF5027; SEIKO NPC Corporation: Tokyo, Japan, February 2010. Available online: http://www.npc.co.jp/en/products/xtal/clock-oscillator/5027-series/ (accessed on 11 May 2018).

48. Meyer, D.G. A Test Set for the Accurate Measurement of Phase Noise on High-Quality Signal Sources. *IEEE Trans. Instrum. Meas.* **1970**, *19*, 215–227. [CrossRef]

49. Barnes, J.A.; Mockler, R.C. The Power Spectrum and Its Importance in Precise Frequency Measurements. *IRE Trans. Instrum.* **1960**, *I-9*, 149–155. [CrossRef]

micromachines

MDPI

Article

Micro-Fabricated Resonator Based on Inscribing a Meandered-Line Coupling Capacitor in an Air-Bridged Circular Spiral Inductor

Eun Seong Kim and Nam Young Kim *

Radio Frequency Integrated Centre (RFIC), Kwangwoon University, Kwangwoon-ro, Nowon-gu, Seoul 01897, Korea; esk@kw.ac.kr
* Correspondence: nykim@kw.ac.kr, Tel.: +82-010-5532-5071

Received: 16 April 2018; Accepted: 8 June 2018; Published: 12 June 2018

Abstract: This letter presents a high-performance micro-fabricated resonator based on inscribing a meandered-line square coupling capacitor in an air-bridged circular spiral inductor on the GaAs-integrated passive device (IPD) technology. The main advantages of the proposed method, which inserts a highly effective coupling capacitor between the two halves of a circular spiral inductor, are the miniaturized size, enhanced coupling coefficient, and improved selectivity. Moreover, using an air-bridge structure utilizes the enhanced mutual inductance in which it maximizes the self-inductance by a stacking inductor layout to obtain a high coupling effect. The simulated and measured S-parameters of a prototype resonator with an effective overall circuit size of 1000 μm × 800 μm are in good agreement. The measured insertion and return losses of 0.41 and 24.21 dB, respectively, at a measured central frequency of 1.627 GHz, as well as an upper band transmission zero with a suppression level of 38.7 dB, indicate the excellent selectivity of the developed resonator.

Keywords: air-bridge; bandpass filter; meandered-line coupling capacitor; micro-fabricated; spiral inductor

---

1. Introduction

Global positioning system (GPS) receivers are being miniaturized and their cost is being minimized to make them accessible for virtually everyone. This has tremendously increased the demand for high-performance compact resonators. An increased growth in the publication rate on microstrip lines and printed circuit board (PCB) technology-based resonators partially reflects the intense interest that resonators for GPS receivers have been generating [1–3]. However, PCB-technology-based resonators still suffer from many bottlenecks, such as limited design flexibility, because of the impracticalities in implementing three-dimensional structures, large critical dimensions (~0.2 mm), and large fabrication tolerances. Therefore, they cannot fulfil the demands of high-level circuit miniaturization. On the other hand, integrated passive device (IPD) technology provides noticeably improved critical dimensions (~5 μm), extended design flexibility with air-bridge structures, low fabrication tolerances, and the capability to integrate with active circuits.

Therefore, IPD-based implementation of high-performance, miniaturized resonators has recently garnered an increased amount of interest from researchers [4–6]. IPDs typically combine a number of passive components into a single package. These devices are increasingly being built using GaAs thin-film substrates rather than on silicon-like semiconductor devices or ceramic, which have very poor characteristics or can be difficult to tune. The integration of a large number of passive components including the individual passive devices such as transmission lines, inductors, and capacitors, or functional passive devices with low loss and minimal crosstalk, is as important as the advancement of active transistor technology [7,8]. IPDs are 50% smaller, 70% thinner, and require 50% less printed circuit board (PCB) area compared to discrete passive devices. In addition, IPDs are a bridge platform between the front and back ends of the semiconductor process. The use of air-bridged differential transformers to miniaturize the effective size of the resonators has been gaining much attention owing to the high self-inductance and increased mutual inductance of such transformers [9]. Metal–insulator–metal (MIM) capacitors with optimized plate areas can be embedded between the two halves of the differential transformers to obtain highly miniaturized, high selectivity resonators [10,11]. However, the requirement of a SiNx dielectric layer increases the fabrication complexity and cost. Other manufacturing challenges include materials and process standardization, library component development, correlation between the design and IPD performance, and integration of the IPD. Also, a higher capacitor density at a lower manufacturing cost is critical [12].

This study reports on the development of a compact resonator based on a micro-fabricated circular spiral inductor and meandered-line coupling capacitor. The noticeable advantages of size miniaturization and selectivity improvement, which are attributed to the coupling capacitor and verified by the S-parameter and derived-lumped-parameter–based analyses, validate the importance of the proposed method in developing high-performance resonators. The measured excellent selectivity of the developed resonator's insertion and return losses of 0.41 and 24.21 dB, respectively, at a measured central frequency of 1.627 GHz, as well as an upper band transmission zero with a suppression level of 38.7 dB, shows the excellent selectivity in the developed resonator.

2. Materials and Methods

Resonator Design

A planar, square, meandered-line coupling capacitor was inscribed in a circular spiral inductor to produce the proposed resonator. The coupling capacitor was inserted in series between the two symmetric halves of the spiral inductor. Therefore, the produced resonator is equivalent to a series combination of its distributed parameters including the loss resistance (R), net inductance (L_{CS}), and coupling capacitance (C_C) [13]. Consequently, the resonator generates a central frequency (f_0) expressed by the following equation:

$$f_0 = \frac{1}{2\pi}\sqrt{\frac{1}{L_{CS}C_C} - \left(\frac{R}{L_{CS}}\right)^2} \tag{1}$$

The proposed resonator based on a micro-fabricated air-bridged circular spiral inductor and meandered-line coupling capacitor is shown Figure 1. The 3D layout and equivalent circuit of the resonator (a), a scanning electron microscopy (SEM) image of the fabricated resonator (b), and an enlarged focused ion beam (FIB) image of an air-bridge structure is shown in (c). In the (a) image, GaAs substrate (400 μm) was used to allow high-speed microelectronics. The first passivation layer deposited was SiNx at 2000 Å thickness. Next, a seed metal layer of Ti/Au at 20/80 nm thickness was deposited. Lastly, Cu at a thickness of 3.35 μm was formed. According to the expression based on a current-sheet expression [14], the inductance of a planar circular spiral inductor can be expressed as follows:

$$L_{CS} = \frac{\mu_0 n^2 d_{avg} c_1}{2}\left(1n(c_2/\rho) + c_3\rho + c_4\rho^2\right) \tag{2}$$

where μ_0 (= $4\pi \times 10^{-7}$ H/m) denotes the space permeability and n is the number of turns. c_1, c_2, c_3, and c_4 are layout-dependent coefficients and exhibit values of 1, 2.46, 0, and 0.2, respectively, for a circular layout. $\rho = (d_{out} - d_{in})/(d_{out} + d_{in})$ and $d_{avg} = (d_{out} + d_{in})/2$ denote the fill ratio and average diameter, respectively, in terms of the outer diameter (d_{out}) and inner diameter (d_{in}). The corresponding metal traces were crossed over using the air-bridge structure, to utilize the enhanced mutual inductance, and the lower metal layer was used as a bridge. A stacked inductor layout was employed to maximize the self-inductance and achieve high area efficiency for a high coupling coefficient [15].

The net coupling capacitance (C_C) between the coupled lines can be expressed as follows:

$$C_C = \left[\varepsilon_0 \left(\frac{1+\varepsilon_s}{2} \right) \frac{K\left(\sqrt{1-k^2}\right)}{K(k)} + \varepsilon_0 \frac{t}{a} \right] L_C \tag{3}$$

where ε_0 and ε_s represent the dielectric constants of air and the substrate, respectively. L_C denotes the total length of the coupled lines and k ($= t/a$) and $K(k)$ represent the elliptical integral of the first kind.

$$Q = \frac{f_0}{B_{3\text{-}dB}} : B_{3\text{-}dB} = f_h - f_l \tag{4}$$

(Q) is the quality factor and (f_0) is the center frequency. ($B_{3\text{-}dB}$) is the 3-dB bandwidth. The meandered-line coupling structure is used to maximize the coupling length and, therefore, the coupling capacitance in a minimum area [16].

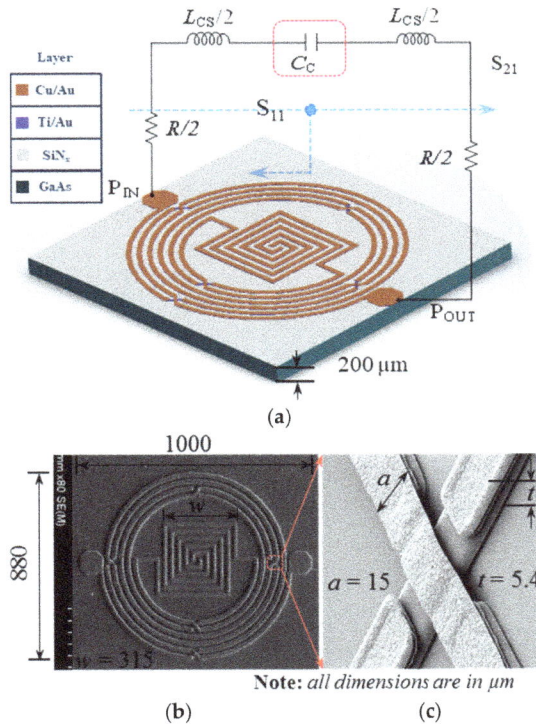

Figure 1. Proposed resonator based on a micro-fabricated air-bridged circular spiral inductor and meandered-line coupling capacitor. (**a**) 3D layout and equivalent lumped-element circuit; (**b**) scanning electron microscopy (SEM) image of fabricated resonator; and (**c**) enlarged focused ion beam (FIB) image of the air-bridge structure.

3. Results and Discussion

A typical resonator based on the proposed design layout was simulated and optimized using Agilent Advanced Design System (ADS) software (version 2016.01, Keysight Technologies, Inc., Santa Rosa, CA, USA) to generate a central frequency of 1.574 GHz. The optimized dimensions were d_{in} = 530 μm, d_{out} = 880 μm, and L_C = 1850 μm. The air-bridged circular inductor consisted of five turns (n = 5) with a 15 μm gap between the corresponding turns. The simulated S-parameters, which are shown in Figure 2a, indicate that the air-bridged inductor alone resonated at a central frequency of 1.98 GHz and exhibited a 5 GHz transmission zero with 22.9 dB suppression. The capacitive effect required that the resonance be provided by the coupling between the top metal layer and the ground aluminum box through the sandwiched GaAs substrate, and has been shown in the detailed equivalent circuit in our previous work [17]. In Figure 2a–d, the spiral inductor graph indicates a circular spiral inductor that is outside of the design and a bandpass filter (BPF) in the inner meandered-line square capacitor. Additionally, the inscribing of the meandered-line coupling capacitor shifted the central frequency downward to 1.574 GHz and, therefore, reduced the effective size of the resonator by 26.67%.

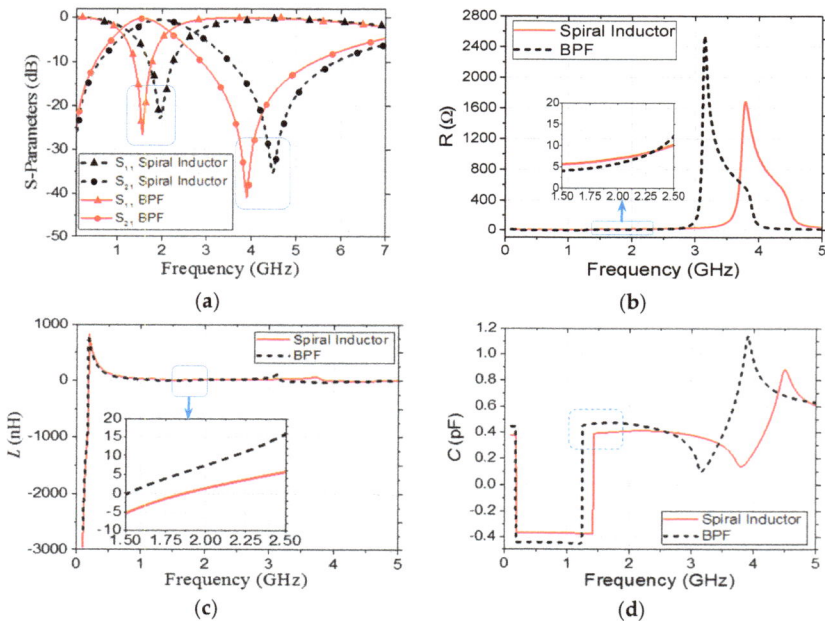

Figure 2. S-parameters and lumped parameters to study the effect of a coupling capacitor. (a) Simulated S11 and S21; (b) resistance (R); (c) inductance (L); and (d) capacitance (C).

Moreover, it markedly enhanced the passband return loss and transmission zero suppression level and, therefore, improved the resonator selectivity. The distributed resistance (R), inductance (L_{CS}), and capacitance (C_C) were extracted from the simulated S-parameters [18] to analyze the effect of the coupling capacitor on the resonator performance, and the corresponding results are shown in Figure 2. It was observed that the use of the meandered-line structure increased L_{CS} (from 1.09 nH to 4.21 nH) and C_C (from 0.38 pH to 0.47 pH) slightly and decreased R (from 5.57 Ω to 4.22 Ω) at 1.574 GHz. In addition, the calculated coupling coefficient (k), which is shown in Figure 3, indicated that the use of the coupling capacitor generated an over-coupled (k = 1.04) resonator at 1.574 GHz. A prototype resonator with optimized dimensions was fabricated on a conventional 6-inch,

200-µm-thick GaAs substrate with a dielectric constant ε_s = 12.85 and loss tangent tan δ = 0.006; the detailed fabrication process is explained elsewhere [19,20]. The coupling capacitor was realized with k = 0.36 and K = 1.6257. Figure 1b,c illustrates a magnified scanning electron microscopy (SEM) image of the fabricated resonator with overall physical dimensions of 1000 µm × 880 µm and a focused ion beam (FIB) image of the air-bridge structure, respectively [21]. Figure 4 shows that the PCB board was attached to a 2 cm² Al Box as a ground to decrease the noise. Both sides were connected to an subminiature version A (SMA) connector to measure with a vector network analyzer (VNA).

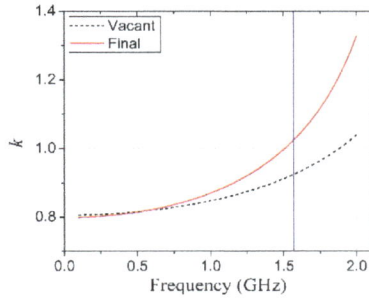

Figure 3. Variation of the coupling coefficient (k) with frequency for an air-bridged spiral inductor alone and for the proposed resonator.

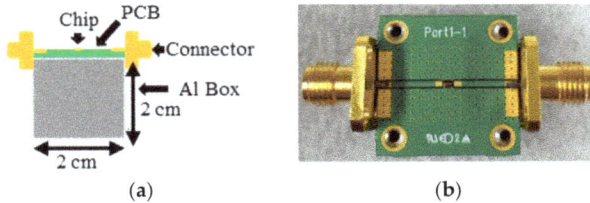

Figure 4. A printed circuit board (PCB) and ground are shown in (**a**) cross-section view; (**b**) and a top section with a design chip.

The S-parameters measured using an Agilent 8510C VNA demonstrated good agreement with the simulated values. However, a small (53 MHz) downward shift of the measured central frequency (1.627 GHz) was observed with respect to the simulated value [22]. The measured 3 dB fractional bandwidth of the passband was 54.08%. The small differences in the central frequency and fractional bandwidth could be because of the substrate dielectric loss, dispersion loss at the inductor bends, and the accuracy of the physical dimensions. Figure 5 shows that the measured insertion and return losses of the passband were 0.41 and 24.21 dB, respectively. A transmission zero appeared at 3.9 GHz with a high suppression level of 38.7 dB. The distributed parameters at 1.627 GHz, extracted from the measured S-parameters, were R = 4.45 Ω, L_{CS} = 1.08 nH, and C_C = 0.48 pF [23].

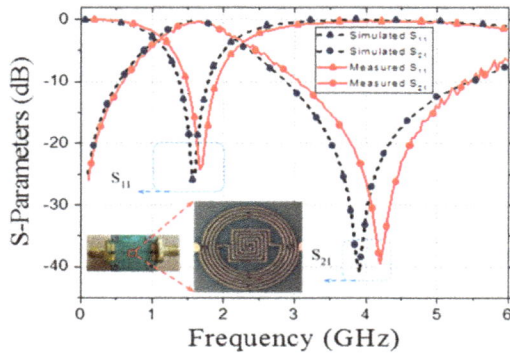

Figure 5. Simulated and measured S-parameters and the pictures of the fabricated resonator.

Table 1, which compares our developed resonator with several recently reported IPD resonators, indicates that our work demonstrates a high-performance resonator with a more compact size and better selectivity owing to the lower insertion loss and higher return loss. In addition, the smaller number of metal layers reduces the fabrication complexity and cost of the device.

Table 1. Performance Comparison with Reported Resonators.

Reference	Technology	CF [1] (GHz)	IL [2] (dB)	RL [3] (dB)	Size (mm^2)	Metal Layers
[17]	GaAs IPD	2.27	0.8	26.1	0.9	2
[24]	Silicon IPD	2.45	2.2	30	1.5	3
[25]	GaAs IPD	7.7	1.63	40.1	4.98	1
This work	GaAs IPD	1.627	0.4	24.2	0.88	2

[1] Central Frequency, [2] Insertion Loss, [3] Return Loss.

4. Conclusions

A highly miniaturized resonator based on an air-bridged circular spiral inductor and an inscribed square meandered-line coupling capacitor was developed using GaAs-based micro-fabrication IPD technology. Lumped-parameter-based analysis revealed that a system of meandered coupled lines between the two halves of a circular spiral inductor could provide high coupling capacitance, miniaturize the effective size, and improve the selectivity of the resulting resonator, significantly. The measured excellent selectivity of the developed resonator's insertion and return losses of 0.41 and 24.21 dB, respectively, at a measured central frequency of 1.627 GHz, as well as an upper band transmission zero with a suppression level of 38.7 dB, shows a good candidate for use in the next upgrade of GPS applications.

Author Contributions: Conceptualization, E.S.K. and N.Y.K.; Methodology, E.S.K. and N.Y.K.; Software, E.S.K.; Validation, E.S.K. and N.Y.K.; Formal Analysis, E.S.K.; Investigation, E.S.K. and N.Y.K.; Resources, E.S.K.; Data Curation, N.Y.K.; Writing—Original Draft Preparation, E.S.K.; Writing—Review & Editing, N.Y.K.; Visualization, N.Y.K.; Supervision, N.Y.K.; Project Administration, N.Y.K.; Funding Acquisition, N.Y.K.

Funding: This research was supported by the Basic Science Research Program through the National Research Foundation of Korea (NRF) funded by the Ministry of Science, ICT and Future Planning (No. 2011-0030079) and a grant supported by the Korean Government (MEST) No. 2015R1D1A1A09057081. This work was also supported by a Research Grant from Kwangwoon University in 2018. Additionally, this research was also supported by the Korean Government (MEST) No. 2016K000117.

Conflicts of Interest: The authors declare no conflict of interest.

References

1. Cheng, C.M.; Yang, C.F. Develop quad-band (1.57/2.45/3.5/5.2 GHz) bandpass filters on the ceramic substrate. *IEEE Microw. Wirel. Compon. Lett.* **2010**, *20*, 268–270. [CrossRef]
2. Rinaldi, M.; Zuniga, C.; Zuo, C.; Piazza, G. Super-high-frequency two-port AlN contour-mode resonators for RF applications. *IEEE Trans. Ultrason. Ferroelectr. Freq. Control* **2010**, *57*, 38–45. [CrossRef] [PubMed]
3. Li, Y.; Wang, C.; Kim, N.Y. A compact dual-band bandpass filter with high design flexibility using fully isolated coupling paths. *Microw. Opt. Technol. Lett.* **2014**, *56*, 642–646. [CrossRef]
4. Wang, C.; Lee, W.S.; Zhang, F.; Kim, N.Y. A novel method for the fabrication of integrated passive device on SI-GaAs substrate. *Int. J. Adv. Manuf. Technol.* **2011**, *52*, 1011–1018. [CrossRef]
5. Wang, C.; Kim, N.Y. Analytical optimization of high-performance and high-yield spiral inductor in integrated passive device technology. *Microelectron. J.* **2012**, *43*, 176–181. [CrossRef]
6. Lin, C.-M.; Chen, Y.-Y.; Felmetsger, V.V.; Senesky, D.G.; Pisano, A.P. AlN/3C–SiC composite plate enabling high-frequency and high-Q micromechanical resonators. *Adv. Mater.* **2012**, *24*, 2722–2727. [CrossRef] [PubMed]
7. Piazza, G.; Stephanou, P.J.; Pisano, A.P. Piezoelectric aluminum nitride vibrating contour-mode MEMS resonators. *J. Microelectromech. Syst.* **2006**, *15*, 1406–1418. [CrossRef]
8. Nam, C.M.; Jung, I.H. High performance RF integrated circuits using the Silicon based RF integrated passive device (RFIPD). In Proceedings of the 2005 Fifth International Conference on Information, Communications and Signal Processing, Bangkok, Thailand, 6–9 December 2005; pp. 1357–1361. [CrossRef]
9. Jeon, H.; Kim, N.Y. A compact quadrature coupler on GaAs IPD process for LTE applications. *IEICE Electron. Express* **2013**, *10*, 1–5. [CrossRef]
10. Li, Y.; Wang, C.; Kim, N.Y. Design of very compact bandpass filters based on differential transformers. *IEEE Microw. Wirel. Compon. Lett.* **2015**, *25*, 439–441. [CrossRef]
11. Lin, C.-M.; Yen, T.-T.; Felmetsger, V.V.; Hopcroft, M.A.; Kuypers, J.H.; Pisano, A.P. Thermally compensated aluminum nitride Lamb wave resonators for high temperature applications. *Appl. Phys. Lett.* **2010**, *97*, 083501. [CrossRef]
12. Qiang, T.; Wang, C.; Kim, N.Y. A compact high-reliability high-performance 900-MHz WPD using GaAs-IPD technology. *IEEE Microw. Wirel. Compon. Lett.* **2016**, *26*, 498–500. [CrossRef]
13. Bahl, I.J. *Lumped Elements for RF and Microwave Circuits*; Artech House, Inc.: Norwood, MA, USA, 2003; ISBN 1-58053-309-4.
14. Mohan, S.S.; del Mar Hershenson, M.; Boyd, S.P.; Lee, T.H. Simple accurate expressions for planar spiral inductances. *IEEE J. Solid-State Circuits* **1999**, *34*, 1419–1424. [CrossRef]
15. Luong, H.C.; Yin, J. *Transformer-Based Design Techniques for Oscillators and Frequency Dividers*; Springer International Publishing: Basel, Switzerland, 2016; pp. 9–12.
16. Huang, C.H.; Chen, C.H.; Horng, T.S. Compact bandpass filter using novel transformer-based coupled filters on integrated passive device glass substrate. *Microw. Opt. Technol. Lett.* **2012**, *54*, 257–262. [CrossRef]
17. Chuluunbaatar, Z.; Adhikari, K.K.; Wang, C.; Kim, N.Y. Micro-fabricated bandpass filter using intertwined spiral inductor and interdigital capacitor. *Electron. Lett.* **2014**, *50*, 1296–1297. [CrossRef]
18. Kim, W.K.; Jung, Y.M.; Cho, J.H.; Kang, J.Y.; Oh, J.Y.; Kang, H.; Lee, H.J.; Kim, J.H.; Lee, S.; Shin, H.J.; et al. Radio frequency characteristics of graphene oxide. *Appl. Phys. Lett.* **2010**, *97*, 193103. [CrossRef]
19. Kim, N.Y.; Adhikari, K.K.; Dhakal, R.; Chuluunbaatar, Z.; Wang, C.; Kim, E.S. Rapid, sensitive, and reusable detection of glucose by a robust radiofrequency integrated passive device biosensor chip. *Sci. Rep.* **2015**, *67*, 7807. [CrossRef] [PubMed]
20. Chen, C.H.; Shih, C.S.; Horng, T.S.; Wu, S.M. Very miniature dual-band and dual-mode bandpass filter designs on an integrated passive device chip. *Prog. Electromagn. Res. Lett.* **2011**, *119*, 461–476. [CrossRef]
21. Wang, C.; Lee, J.H.; Kim, N.Y. High-performance integrated passive technology by advanced SI-GaAs-based fabrication for RF and microwave applications. *Microw. Opt. Technol. Lett.* **2010**, *52*, 618–623. [CrossRef]
22. Kim, N.K.; Dhakal, R.; Adhikari, K.K.; Wang, S.; Kim, E.S. A reusable robust RF biosensor using microwave resonator by integrated passive device technology for quantitative detection of glucose level. *Biosens. Bioelectron.* **2015**, *67*, 687–693. [CrossRef] [PubMed]
23. Li, Y.; Wang, C.; Yao, Z.; Kim, N.Y. Very compact differential transformer-type bandpass filter with mixed coupled topology using integrated passive device technology. *Microelectron. J.* **2015**, *46*, 1459–1463. [CrossRef]

Micromachines **2018**, *9*, 294

24. Liu, K.; Frye, R.; Emigh, R. Bandpass filter with balun function from IPD technology. In Proceedings of the Electronic Components and Technology Conference, Lake Buena Vista, FL, USA, 27–30 May 2008; pp. 718–723.

25. Chuluunbaatar, Z.; Wang, C.; Kim, N.Y. Internally loaded cross-coupled open-loop filters for a miniaturized bandpass filter using integrated passive device technology. *Microw. Opt. Technol. Lett.* **2014**, *56*, 2737–2740. [CrossRef]

micromachines

MDPI

Article

Wide Acoustic Bandgap Solid Disk-Shaped Phononic Crystal Anchoring Boundaries for Enhancing Quality Factor in AlN-on-Si MEMS Resonators

Muhammad Wajih Ullah Siddiqi [1,*] and Joshua E.-Y. Lee [1,2]

[1]   Department of Electronic Engineering, City University of Hong Kong, Kowloon, Hong Kong, China;
      josh.lee@cityu.edu.hk
[2]   State Key Laboratory of Millimeter Waves, City University of Hong Kong, Kowloon, Hong Kong, China
*   Correspondence: musiddiqi2-c@my.cityu.edu.hk; Tel.: +852-6488-7452

Received: 25 May 2018; Accepted: 14 August 2018; Published: 18 August 2018

Abstract: This paper demonstrates the four fold enhancement in quality factor (Q) of a very high frequency (VHF) band piezoelectric Aluminum Nitride (AlN) on Silicon (Si) Lamb mode resonator by applying a unique wide acoustic bandgap (ABG) phononic crystal (PnC) at the anchoring boundaries of the resonator. The PnC unit cell topology, based on a solid disk, is characterized by a wide ABG of 120 MHz around a center frequency of 144.7 MHz from the experiments. The resulting wide ABG described in this work allows for greater enhancement in Q compared to previously reported PnC cell topologies characterized by narrower ABGs. The effect of geometrical variations to the proposed PnC cells on their corresponding ABGs are described through simulations and validated by transmission measurements of fabricated delay lines that incorporate these solid disk PnCs. Experiments demonstrate that widening the ABG associated with the PnC described herein provides for higher Q.

Keywords: microelectromechanical systems (MEMS); AlN-on-Si resonators; phononic crystal; anchor loss; quality factor; acoustic bandgap

---

1. Introduction

Quartz crystal resonators are integral components in existing radio frequency (RF) communication systems. With the advantages of small form factor and CMOS process compatibility, microelectromechanical systems (MEMS) resonators have been of interest for providing integrated clock-chip solutions. Among MEMS resonators, piezoelectric-on-silicon resonators have the advantage of possessing more efficient electromechanical coupling (in relation to capacitive silicon resonators), lower acoustic loss and high power handling capacity (in relation to piezoelectric film body resonators) [1]. High Q and efficient coupling lead to lower motional resistance (R_m). As high Q and low R_m are desired for low phase noise oscillators [2], various approaches have been proposed to improve the Qs (thus also reducing R_m) of different piezoelectric resonators by suppressing anchor losses. This includes using biconvex plates to trap acoustic energy in the center of the acoustic cavity and thus minimizing the distribution of energy closed to the anchors [3]. For AlN-body resonators, butterfly-shaped AlN plates have been proposed [4]. Another approach has been to etch acoustic reflectors into the anchoring boundaries of the resonator to reflect the outgoing acoustic waves from the supports back into the resonator [5]. Moreover, phononic crystals (PnCs) have been hybridized with Silicon (Si) [6], Aluminum Nitride (AlN) [7,8], AlN-on-Si [9–12] and Gallium Nitride (GaN) [13] resonators with either one-dimensional (1D) or two-dimensional (2D) periodicity. PnCs are inhomogeneous periodic structures (e.g., solid–air) that shape the transmission of phonons through the PnC. This ability to shape the behavior of acoustic waves in the PnC is due to the generation of an ABG that arises from the periodic alternation of acoustic properties (velocity and density) in the PnC. From the literature, PnCs of different shapes and geometries have been investigated in relation to their effect on the width of the associated ABG. The more common PnC shapes proposed in the literature include air-holes in substrate [11,14–19], rings [9,10], cross-shaped [7,20], snowflake air inclusions [21] and fractals [22]. When a defect is introduced in a periodic PnC structure, various resonators and waveguides may be formed. PnCs can be engineered with a desired bandgap around a target center frequency and can be incorporated either in 1D or 2D patterns to realize cavity resonators and waveguides. The PnCs around the cavity behave as an acoustic shield that confines the acoustic energy within the cavity. However, cavity resonators and waveguides are associated with lower Qs and higher insertion losses. Alternatively, PnC based supporting tethers [6–11] have been used in lieu of simple beam tethers in resonators to reduce anchor loss and thus enhancing Q. As propagation of acoustic waves at frequencies within the ABG is prohibited within the PnC [23], placing PnCs close to the resonator where acoustic waves are expected propagate out has the benefit of reducing anchor loss. On this note, the size of the ABG depends on the shape of the PnC unit cell.

Using simulations based on AlN as the solid matrix, Ardito [24] showed that shaping the PnC unit cell as a solid-disk produces a notably wider ABG compared to other more common shapes. This interesting PnC design resulted from an optimization process through applying the Bidirectional Evolutionary Structural Optimization (BESO) technique. Recently, we applied the solid-disk PnC unit cell topology proposed by Ardito [24] to a piezoelectric AlN-on-Si Lamb mode resonator to demonstrate the proof of concept [25].

Most reports of the use of PnCs to enhance Q of piezoelectric resonators have been on AlN-body resonators where the enhanced Qs have been rather limited [7,8]. While there has been some coverage of PnCs applied to AlN-on-Si resonators, the enhanced Qs have been limited to below 4000 [9–12]. In comparison, we have previously demonstrated Qs around 10,000 using biconvex resonators to reduce anchor loss to the point where electrode-related losses become dominant [26] but not with PnCs to date [27]. This work seeks to validate the effectiveness of the uniquely optimized PnC design proposed by Ardito [24] in reaching similar levels of Q. We show that these PnCs are able to enhance Qs to the same levels as the biconvex resonators reported in [26]. These PnC enhanced resonators also show similar Qs with respect to the number of electrodes, thus verifying that the PnCs can likewise reduce anchor to a point where other kinds of losses related to the electrodes become dominant. However, compared to the biconvex resonators, the use of PnCs anchors avoids the need to modify the resonator shape and compromise on the coupling area.

This work extends from the preliminary work reported in [25] by experimentally validating the existence of a wide band gap. By incorporating the PnC into a delay line, we demonstrate a stop band that 120 MHz wide around a center frequency of 141 MHz. Furthermore, we demonstrate how adjustments to the dimensions of the PnC cell affect the size of the band gap. These wide ABG PnCs were applied as the anchoring boundaries of 7th-order mode AlN-on-Si Lamb mode resonators with resonant frequencies of around 144 MHz to demonstrate enhancement in Q of over 4-fold.

2. Design and Simulations of ABG

As depicted by Figure 1a, the basic PnC unit cell has a lattice constant, a = 22 μm and comprises a solid disk with a radius, r = 8 μm, with a thickness, t = 10 μm (set by the fabrication process [28]). The solid disk in a given unit cell is linked to the solid disks in adjacent cells by slender links. The minimum width of these links, w = 2 μm, is limited by the fabrication process. Using COMSOL Multiphysics (version 5.2a, COMSOL Co., Ltd., Shanghai, China) finite element (FE) analysis, the phononic band structure was computed by applying Floquet periodic boundary conditions on the unit cell along the x- and y-axis and performing a parametric sweep in the wave vector k bounding the first Brillouin zone illustrated by Figure 1b. The lattice parameter and disk radius were tuned in the FE simulations to synthesize a wide complete ABG for which the center frequency of the band coincides closely with the resonant frequency of the Lamb mode resonator to be physically bound by the PnC.

The size of the ABG can be expressed in terms of a dimensionless parameter using the gap-to-mid gap ratio to avoid frequency dependence:

$$BG = \frac{fu - f_L}{(fu + f_L)/2} \tag{1}$$

where fu and f_L are the upper and lower bounds, respectively, of the ABG. The widest complete ABG of 82 MHz spanning from 93–175 MHz (depicted in Figure 1c) was obtained when the inter-cell link was set to its minimum (i.e., $w = 2$ μm), which corresponds to a bandgap ratio of 61%. Increasing w while keeping all other geometrical parameters unchanged, the ABG was reduced to 44% when $w = 3$ μm, and further down to 29% when $w = 4$ μm. As such, the simulations show that widening the inter-cell link reduces the ABG. Table 1 compares the simulated ABG of the proposed solid disk PnC with the commonly used air-hole, ring, cross inclusion, square hole and fractal PnCs assuming the same lattice size $a = 22$ μm and minimum feature size of 2 μm. The dimensions of each PnC shape have been chosen to yield close to the widest possible ABG. The center frequencies of the associated ABGs are within a similar range. A more exact alignment of center frequencies would require adjustments to the lattice parameter and various dimensions in each PnC cell, although the ultimate conclusion on which shape provides the widest ABG remains the same. The proposed solid-disk PnC possesses the highest gap-to-mid gap ratio (BG).

Table 1. Comparison of simulated ABGs between various PnC shapes using the same lattice parameter $a = 22$ μm and minimum feature size of 2 μm.

PnC Shape	Dimensions (μm)	ABG Range (MHz)	f_c (MHz)	BG (%)
Air-Hole		136–147	141.5	7.7
Ring		107–144	125.5	29.4
Cross Inclusion		89–147	118	49.1
Square Hole		-	-	-
Fractal		146–192	169	27.2
Solid-Disk (this work)		93–175	134	61.1

Next, FE models to simulate the effect of the PnC (and absence of) on the transmission through a delay line were considered to compare against the experimental results. Figure 2a,b shows the 3D FE simulated displacement profiles of the control delay line having solid slab as transmission medium and delay line with PnCs as transmission medium between drive and sense interdigitated transducers (IDTs) respectively. To reduce computational time, the frequency response simulations were carried out on a single row of 10 solid-disk PnCs with 6 finger electrodes for the IDTs on either side. Periodic boundary conditions were applied in the transverse direction (i.e., y-axis with reference to coordinate system annotated in Figure 2a,b. In the control delay line the PnCs were replaced by a solid slab. To minimize the effect of the reflected waves, Perfectly Matched Layers (PMLs) were introduced at the ends of the delay lines in the longitudinal direction. The FE simulation result illustrates the drop in transmission at a frequency range that corresponds to the associated ABG of the PnC as the PnC transmission medium effectively blocks the propagation of waves generated from the drive IDTs for frequencies that lie within the ABG that otherwise propagate through.

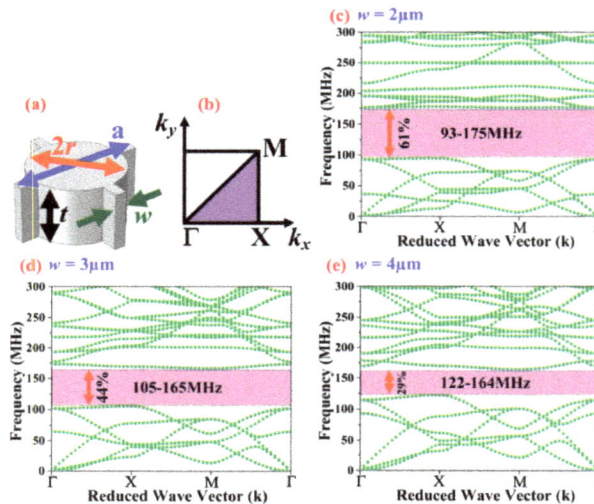

Figure 1. (**a**) Perspective view of solid disk PnC basic unit cell; (**b**) 1st irreducible Brillouin zone in k-space, frequency band diagrams by finite element (FE) analysis of a solid disk PnC for various inter-cell link widths of (**c**) w = 2 µm (**d**) w = 3 µm (**e**) w = 4 µm. Figures modified from the preliminary work in [25].

3. Experimental Validation of ABG

To experimentally verify the existence of the simulated ABGs associated with the solid-disk PnCs, we designed and fabricated three delay lines using a standard AlN-on-SOI MEMS process [28], two of which are depicted in Figure 3. The delay lines comprise a pair of IDTs with a pitch of 22 µm on each side of the delay line and an aperture of 22 µm. In two of the delay lines, 12 rows of solid disk PnCs were etched between the IDTs. Figure 3a shows one of these delay lines with solid disk PnCs in the transmission medium. The PnCs either had inter-cell link widths of w = 2 µm or w = 3 µm. The third delay line, depicted in Figure 3b, serves as a reference device by incorporating just a solid silicon slab as the transmission medium between the IDTs. As shown in the side view schematic in Figure 3c, the regions with IDT electrodes are released for both types of delay lines.

In the experiments, to reduce parasitic feedthrough, a fully-differential probe configuration was applied to the IDTs. The measured transmission (S_{21}) curves from the delay lines with two different PnCs and a solid silicon slab as the propagating medium are shown in Figure 4. From Figure 4, we see that the delay lines incorporating the solid disk PnC both show an abrupt drop in S_{21} in the form of a wide band-stop filter. The drop in S_{21} measured around the middle of the stop band was 30 dB. The experimental transmission response of the PnC delay lines generally agree with the FE simulated response. The difference in the amount of attenuation between experiments and simulation within the stop band may be due to the difference between employing a finite number of rows (in the actual device) as opposed to an infinite lattice (assumed in the FE simulations in relation to the applied periodic boundary conditions). From the experiments, we see that the PnC with the narrower inter-cell links (w = 2 μm) yielded a wider stop band compared to the wider inter-cell link (w = 3 μm).

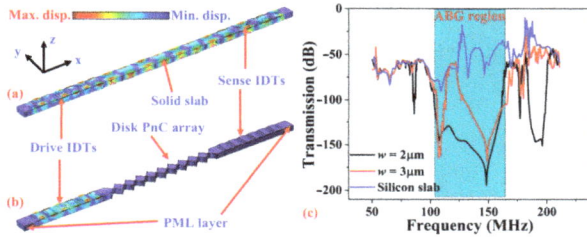

Figure 2. FE simulations with periodic boundary conditions applied along the y-axis—Displacement profiles of the (**a**) reference delay line having solid slab as the transmission medium and (**b**) delay line with PnCs (with inter-cell link width w = 3 μm) as the transmission medium. (**c**) Simulated transmission S_{21} of the three delay lines showing the existence of the bandgap associated with the solid disk PnC observed through significant drops in transmission within a wide stop band.

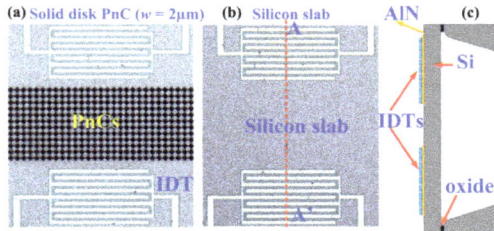

Figure 3. Micrographs of the fabricated delay lines incorporating (**a**) PnCs (with link width of 2 μm) for the transmission medium and (**b**) solid silicon slab as the transmission medium (as the control device for comparison). (**c**) Transverse-view schematic of the delay line showing that the regions with IDT electrodes are released for both types of delay lines.

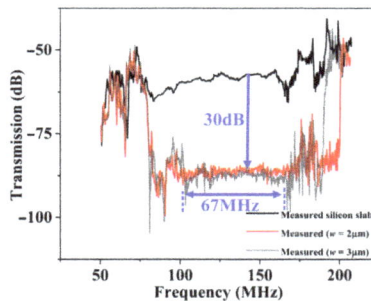

Figure 4. Measured transmission S_{21} of the three delay lines showing the existence of the bandgap associated with the solid disk PnC observed through the 30 dB drop in transmission within a wide stop band.

4. Experimental Validation of Q Enhancement with PnC Anchors

4.1. Resonators with Partial Coverage of Three IDT Fingers.

Having verified the existence of the wide ABG associated with the solid disk PnC, we applied the PnCs as anchoring boundaries to a rectangular plate resonator to demonstrate their effectiveness in enhancing Q by reducing anchor loss. The resonator has a center-to-center electrode pitch W_p = 30 µm (as shown in Figure 5a), designed to be transduced in the 7th-order symmetric Lamb mode that occurs at a frequency of 141 MHz, though lower harmonic modes can be transduced as well (as will be seen in Section 4.3). The resonant frequency for any given harmonic mode is described by:

$$f = \frac{nv}{2W_r} \tag{2}$$

where v is the acoustic velocity of the resonator, W_r is the width of the resonator, and n is the mode number of the respective harmonic. Hence to preferentially transduce the 7th-order Lamb mode, $W_p = W_r/7$. As shown by the micrograph depicted in Figure 5b, the Lamb mode resonator is bound on each anchoring side by solid disk PnCs. As the frequency of the 7th-order symmetric Lamb mode lies well within the ABG of the PnCs, outgoing acoustic waves from the supporting tether are prohibited from propagating through the PnCs. Solid disk PnCs with two different inter-cell link widths were fabricated (w = 2 µm, w = 3 µm) to investigate the effect of the size of the ABG on Q. Figure 5c provides a zoom-in image of the solid disk PnCs for one of the devices in relation to the Lamb mode resonator. The device without PnCs depicted in Figure 5a served as a control device. As these devices were released by trench etching through the bulk substrate, to ensure that the boundary conditions between the control device and the resonator devices incorporating PnCs were similar, the same trench size was used for all devices. The resonators were partially covered with IDTs, specifically only three IDT fingers to reduce the effect of electrode-related losses.

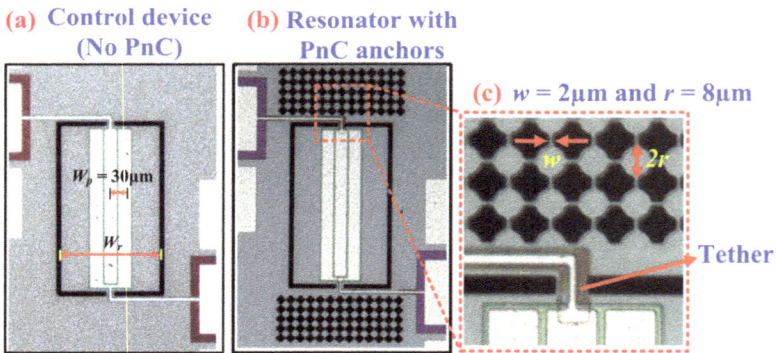

Figure 5. Optical micrographs of the fabricated devices partially covered with 3 IDT fingers: (**a**) Control device with no PnCs, (**b**) PnC bounded resonator, and (**c**) close-up view of the PnC matrix in relation to the Lamb mode resonator connected through the tether whose width was kept to the minimum allowed by the process.

To experimentally validate the effect of the ABG size on the performance of the 7th-order symmetric Lamb mode AlN-on-Si resonators, we fabricated all these devices using the same AlN-on-SOI MEMS process used to fabricate the delay lines described in previous sections. The width of the supporting tether was kept to its practical minimum (16 µm) as allowed by the process with the aim to minimize losses through the tethers particularly when these are wide. As depicted in Figure 5b, five rows of PnCs were employed in the direction of outgoing wave from the tethers.

On this note, it has been previously shown that the effectiveness of the PnCs in enhancing Q plateaus after three PnC cells [9]. Short-open-load-through (SOLT) calibration was performed prior to measure all resonators. Six samples of each of the three resonator types were tested (i.e., a total of 18 samples tested) to ensure repeatability. The measured resonant frequencies of the Lamb mode resonators coincide with the FE simulated values. Figure 6a depicts 7th-order harmonic Lamb mode simulated by FE. Fixed boundary conditions were applied to the tether faces in the yz-plane. Figure 6b shows the corresponding measured S_{21} of a resonator with PnC anchors (w = 2 μm) and a control device (no PnC anchors) to illustrate the increase in Q and corresponding reduction in insertion loss by incorporating the PnCs into the anchors. These results are typical of the measurements carried out over multiple samples tested for repeatability. Figure 7 shows the extracted values of unloaded Q (Q_u) from the S_{21} responses of the 18 samples tested. It is worth pointing out that the values of Q measured for the control resonators are typical of the Lamb wave mode resonators we have fabricated with the same process and tested previously. We have also previously shown that the length of the supporting tether does not significantly alter the value of Q of these resonators based on experiments on multiple samples [29]. As such, the control resonators described herein do not represent particularly poorly designed resonators to yield a sub-optimal Q. As seen from Figure 7, incorporating PnCs into the anchors increases the mean Q_u relative to the control device as much as 3.6-fold. The extent of enhancement in Q_u increases with narrower inter-cell links, a trend that was also observed previously in the case air-hole in substrate PnCs [11]. For reference, the levels of Q attained using PnCs with the narrower links are similar to the 7th-order biconvex resonators of the same frequencies with three IDT fingers described in [26].

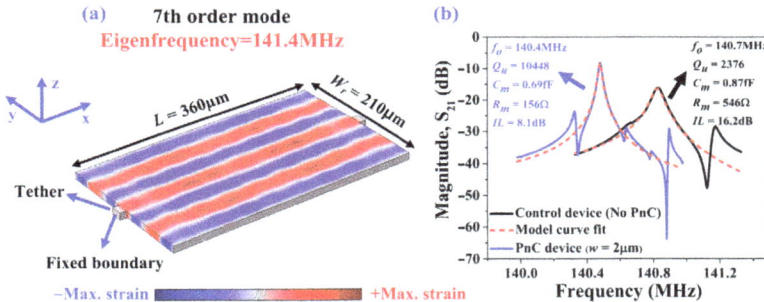

Figure 6. (a) FE simulation of the 7th-order symmetric Lamb mode where the contours denote the associated y-direction strain component. (b) Typical measurements of the transmission S_{21} of two of the partially covered resonators with three IDT fingers with PnC anchors (w = 2 μm) and a control device (no PnC anchors) for comparison to illustrate the increase in Q; the resonators were transduced in the 7th-order symmetric Lamb mode.

4.2. Resonators with Full Coverage of Seven IDT Fingers.

To investigate the effect of the electrodes on the degree of Q enhancement when applying the PnCs in the anchoring boundaries, we designed and fabricated another set of resonators with same size of those in the previous section but with full coverage of IDT fingers (i.e., seven IDT fingers). As in the previous set of resonators with partial coverage of IDT electrodes, we considered four designs: a control device with no PnCs, and three devices with PnCs comprising different inter-cell link widths (w = 2 μm, w = 3 μm, w = 4 μm) to investigate the effect of the size of the ABG on Q in the limit where electrode-related losses are more dominant. Figure 8a depicts a micrograph of the control device with full IDT electrode coverage, and Figure 8b depicts a micrograph of one of the devices with PnCs anchors. Figure 8c provides a zoom-in image of the solid disk PnCs for one of the devices in relation to the Lamb mode resonator with full IDT electrode coverage.

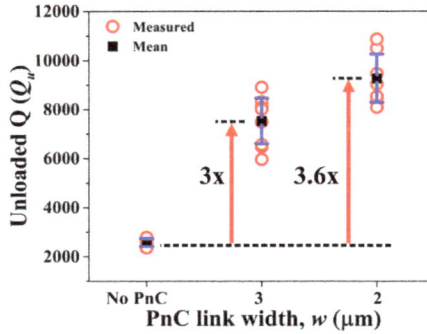

Figure 7. Partially covered resonators with three IDT fingers transduced in the 7th-order symmetric Lamb mode—Extracted values of unloaded Q (Q_u) of the control design with no PnCs in comparison to the two other designs with PnCs in the anchors with different inter-cell link widths. The black squares denote the mean value while the error bars denote the standard deviation over six samples for each resonator design.

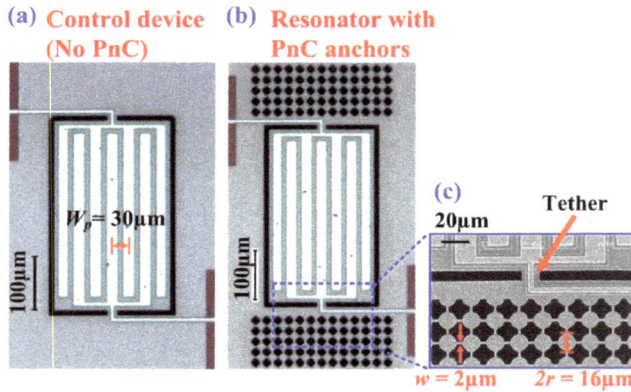

Figure 8. Optical micrographs of the fabricated devices: (**a**) Control device with no PnCs, (**b**) PnC bounded resonator, and (**c**) close-up view of the PnC matrix in relation to the Lamb mode resonator connected through the tether whose width was kept to the minimum allowed by the process.

To experimentally validate the effect of the ABG size on the performance of the 7th-order mode AlN-on-Si resonators with seven IDT fingers, five samples of each of the four resonator types were tested (i.e., total of 20 samples tested) to ensure repeatability. The measured resonant frequencies of the Lamb mode resonators are in agreement with the FE simulated values. Figure 9 depicts the extracted values of unloaded Q (Q_u) from the S_{21} responses of the 20 samples tested. As seen from Figure 9, incorporating PnCs into the anchors increases the mean Q_u relative to the control device by as much as 4.2-fold. As with the case of the resonators partially covered with IDT electrodes, we also see that reducing the PnC link width results in an increase in Q. But we can also see that increasing the number of IDT fingers has reduced the maximum achievable level of Q. For reference, the levels of Q attained using PnCs with the narrowest links are similar to the 7th-order biconvex resonators of the same frequencies with 5 IDT fingers described in [26]. As such, the levels of Q and trends of the resonators bounded by the widest ABG PnCs (i.e., the narrowest links) are what have been observed previously in biconvex resonators. As such, the results reported herein demonstrate that the wide ABG solid-disk PnCs are able to reduce anchor loss to a point whereby electrode-related losses begin to dominate over anchor loss.

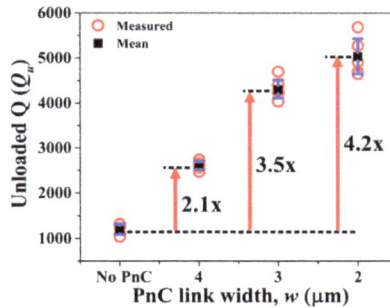

Figure 9. Fully covered resonators with seven IDT fingers transduced in the 7th-order symmetric Lamb mode—Extracted values of unloaded Q (Q_u) of the control design with no PnCs in comparison to the three other devices with PnCs in the anchors with different inter-cell link widths. The black squares denote the mean value while the error bars denote the standard deviation over five samples.

4.3. Frequency Selectivity of Q Enhancement.

To show that the effect of Q enhancement applies to frequencies that lie within the ABG, we tested the very same resonators at their fundamental mode (20.3 MHz) and 3rd-order mode (61 MHz). Both these frequencies lie outside the theoretical bandgap of the same PnC topology but with different inter-cell link widths. We tested four die samples for each of the four designs (i.e., 16 devices in total). Figure 10a depicts the fundamental Lamb mode simulated by FE (occurring around 20.3 MHz), while Figure 10b summarizes the associated unloaded Qs extracted from the measurements. Similarly, Figure 11a depicts the 3rd-order Lamb mode simulated by FE (occurring around 61 MHz), while Figure 11b summarizes the associated unloaded Qs extracted from the measurements of the very same devices. We see that, for either harmonic mode, the PnCs do not provide any enhancement of Q.

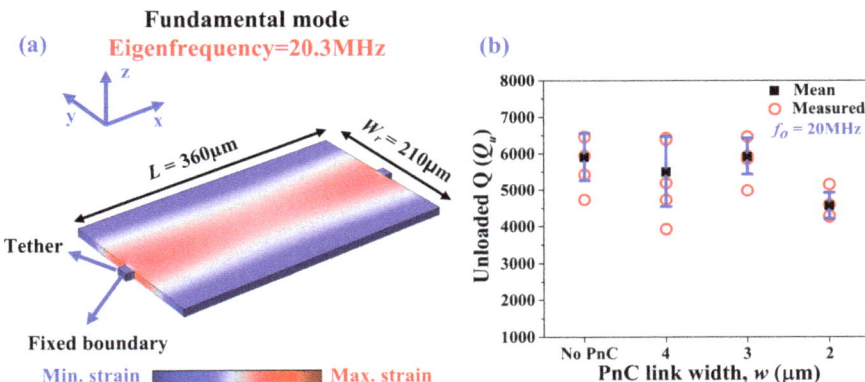

Figure 10. (a) FE simulation of the fundamental Lamb mode (occurring around 20.3 MHz) where the contours denote the associated y-direction strain component. (b) Fully covered resonators with seven IDT fingers—Extracted values of unloaded Q (Q_u) of the control design with no PnCs compared to the other three designs with PnCs in the anchors with different inter-cell link widths transduced at the fundamental Lamb mode (20.3 MHz), which lies outside the ABG. The black squares denote the mean value, while the error bars denote the standard deviation over four samples.

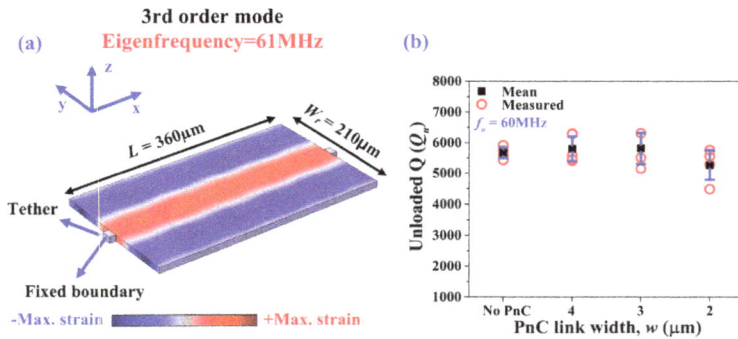

Figure 11. (**a**) FE simulation of the 3rd-order symmetric Lamb mode (occurring around 61 MHz), where the contours denote the associated y-direction strain component. (**b**) Fully covered resonators with seven IDT fingers—Extracted values of unloaded Q (Q_u) of the control design with no PnCs compared to the other three designs with PnCs in the anchors with different inter-cell link widths transduced at the 3rd-order symmetric Lamb mode around 61 MHz that lies outside the ABG. The black squares denote the mean value while the error bars denote the standard deviation over four samples.

Table 2 compares the performance of the PnC resonator hybrids (3 IDTs and inter-cell link widths of 2 μm) transduced in the 7th-order symmetric Lamb mode reported herein with other state-of-the-art piezoelectric AlN and AlN-on-Si resonators disclosed in the literature for similar resonant frequencies.

Table 2. Performance comparison of proposed PnC/resonator hybrids with the state-of-the-art piezoelectric AlN and AlN-on-Si resonators that share similar resonant frequencies.

Reference	Technology	Frequency (MHz)	Q	$f \cdot Q$ (10^{11} Hz)
[30]	AlN	175	1500	2.6
[31]	AlN	101.71	1257	1.27
[2]	AlN-on-Si	106	4000	4.24
[27]	AlN-on-Si	100	5369	5.36
[10]	AlN-on-Si	178	1400	2.49
[9]	AlN-on-Si	141.5	5730	8.10
This work (w/o PnC)	AlN-on-Si	140.9	2510	3.53
This work (with PnC)	AlN-on-Si	140.5	10,492	14.7

5. Conclusions

In conclusion, we have experimentally demonstrated the existence of a wide ABG associated with a solid-disk PnC unit cell topology. By incorporating these PnCs into a delay line, we have experimentally demonstrated a bandgap ratio of 85%, inferred from the transmission measurement of a delay line incorporating such PnCs. Moreover, we have shown that the geometrical dependence of the bandgap ratio predicted by FE simulations is corroborated by the experiments. We have applied these solid-disk wide ABG PnCs to the anchors of AlN-on-Si Lamb mode resonators to demonstrate enhancements in Q by fourfold. The geometric dependence of the ABG is also evident in the extent of enhancement in Q provided by the PnC.

Author Contributions: Supervision, J.E.-Y.L.; Validation, M.W.U.S.; Writing—original draft, M.W.U.S.; Writing—review & editing, J.E.-Y.L.

Funding: This work was supported by a grant from the Research Grant Council of Hong Kong, University Grants Committee under project number CityU 11206115.

Conflicts of Interest: The authors declare no conflict of interest.

References

1. Ho, G.K.; Abdolvand, R.; Sivapurapu, A.; Humad, S.; Ayazi, F. Piezoelectric-on-silicon lateral bulk acoustic wave micromechanical resonators. *J. Microelectromech. Syst.* **2008**, *17*, 512–520. [CrossRef]
2. Abdolvand, R.; Lavasani, H.M.; Ho, G.K.; Ayazi, F. Thin-film piezoelectric-on-silicon resonators for high frequency reference oscillator applications. *IEEE Trans. Ultrason. Ferroelectr. Freq. Control* **2008**, *55*, 2596–2606. [CrossRef] [PubMed]
3. Lin, C.M.; Lai, Y.J.; Hsu, J.-C.; Seneskey, D.G.; Pisano, A.P. High-Q aluminum nitride Lamb wave resonators with biconvex edges. *Appl. Phys. Lett.* **2011**, *99*, 143501. [CrossRef]
4. Zou, J.; Lin, C.-M.; Tang, G.; Pisano, A.P. High-Q Butterfly-shaped AlN Lamb wave resonators. *IEEE Electron Device Lett.* **2017**, *38*, 1739–1742. [CrossRef]
5. Harrington, B.P.; Abdolvand, R. In-plane acoustic reflectors for reducing effective anchor loss in lateral-extensional MEMS resonators. *J. Micromech. Microeng.* **2011**, *21*, 085021. [CrossRef]
6. Feng, D.; Xu, D.; Wu, G.; Xiong, B.; Wang, Y. Phononic crystal strip based anchors for reducing anchor loss of micromechanical resonators. *J. Appl. Phys.* **2014**, *115*, 024503. [CrossRef]
7. Lin, C.-M.; Hsu, J.-C.; Senesky, D.G.; Pisano, A.P. Anchor loss reduction in AlN Lamb wave resonators using phononic crystal strip tethers. In Proceedings of the 2014 IEEE International Frequency Control Symposium (FCS), Taipei, Taiwan, 19–22 May 2014; pp. 371–375.
8. Wu, G.; Zhu, Y.; Merugu, S.; Wang, N.; Sun, C.; Gu, Y. GHz spurious mode free AlN Lamb wave resonator with high figure of merit using one dimensional phononic crystal tethers. *Appl. Phys. Lett.* **2016**, *109*, 013506. [CrossRef]
9. Zhu, H.; Lee, J.E.-Y. Design of phononic crystal tethers for frequency-selective quality factor enhancement in AlN piezoelectric-on-silicon resonators. *Procedia Eng.* **2015**, *120*, 516–519. [CrossRef]
10. Sorenson, L.; Fu, J.L.; Ayazi, F. One-dimensional linear acoustic bandgap structures for performance enhancement of AlN-on-Silicon micromechanical resonators. In Proceedings of the 16th International Solid-State Sensors, Actuators and Microsystems Conference, Beijing, China, 5–9 June 2011; pp. 918–921.
11. Zhu, H.; Lee, J.E.-Y. AlN piezoelectric on silicon MEMS resonator with boosted Q using planar patterned phononic crystals on anchors. In Proceedings of the 28th IEEE International Conference on Micro Electro Mechanical Systems (MEMS), Estoril, Portugal, 18–22 January 2015; pp. 797–800.
12. Rawat, U.; Nair, D.R.; DasGupta, A. Piezoelectric-on-Silicon array resonators with asymmetric phononic crystal tethering. *J. Microelectromech. Syst.* **2017**, *26*, 773–781. [CrossRef]
13. Wang, S.; popa, L.C.; Weinstein, D. GaN MEMS resonator using a folded phononic crystal structure. In Proceedings of the Hilton Head Workshop 2014: A Solid-State Sensors, Actuators and Microsystems, Sonesta Resort, Hilton Head Island, 8–12 June 2014; pp. 72–75.
14. Wang, N.; Tsai, J.M.-L.; Hsiao, F.-L.; Soon, B.W.; Kwong, D.-L.; Palaniapan, M.; Lee, C. Micromechanical resonators based on silicon two-dimensional phononic crystals of square lattice. *J. Microelectromech. Syst.* **2012**, *21*, 801–810. [CrossRef]
15. Li, F.; Liu, J.; Wu, Y.H. The investigation of point defect modes of phononic crystal for high Q resonance. In Proceedings of the 16th International Solid-State Sensors, Actuators and Microsystems Conference, Beijing, China, 5–9 June 2011; pp. 438–441.
16. Mohammadi, S.; Eftekhar, A.A.; Hunt, W.D.; Adibi, A. High-Q micromechanical resonators in a two-dimensional phononic crystal slab. *Appl. Phys. Lett.* **2009**, *94*, 051906. [CrossRef]
17. Mohammadi, S.; Eftekhar, A.A.; Khelif, A.; Adibi, A. A high-quality factor piezoelectric-on-substrate phononic crystal micromechanical resonator. In Proceedings of the 2009 IEEE International Ultrasonics Symposium, Rome, Italy, 20–23 September 2009; pp. 1158–1160.
18. Huang, C.-Y.; Sun, J.-H.; Wu, T.-T. A two-port ZnO/silicon Lamb wave resonator using phononic crystals. *Appl. Phys. Lett.* **2010**, *97*, 031913. [CrossRef]
19. Wang, N.; Tsai, J.M.; Soon, B.W.; Kwong, D.-L.; Hsiao, F.-L.; Palaniapan, M.; Lee, C. Experimental demonstration of microfabricated phononic crystal resonators based on two-dimensional silicon plate. In Proceedings of the 2011 Defense Science Research Conference and Expo (DSR), Singapore, 3–5 August 2011; pp. 1–4.
20. Lu, R.; Manzaneque, T.; Yang, Y.; Gong, S. Lithium Niobate phononic crystals for tailoring performance of RF laterally vibrating devices. *IEEE Trans. Ultrason. Ferroelectr. Freq. Control* **2018**, *65*, 1–11. [CrossRef] [PubMed]

21. Baboly, M.G.; Reinke, C.M.; Graiffin, B.A.; El-Kady, I.; Leseman, Z.C. Acoustic waveguiding in a silicon carbide phononic crystals at microwave frequencies. *Appl. Phys. Lett.* **2018**, *112*, 103504. [CrossRef]

22. Kuo, N.-K.; Piazza, G. Ultra high frequency air/aluminum nitride fractal phononic crystals. In Proceedings of the 2011 Joint Conference of the IEEE International Frequency Control and the European Frequency and Time Forum (FCS), San Francisco, CA, USA, 2–5 May 2011; pp. 1–4.

23. Benchabane, S.; Khelif, A.; Rauch, J.-Y.; Robert, L.; Laude, V. Evidence for complete surface wave band gap in a piezoelectric phononic crystal. *Phys. Rev. E* **2006**, *73*, 065601. [CrossRef] [PubMed]

24. Ardito, R.; Cremonesi, M.; D'Alessandro, L.; Frangi, A. Application of optimally-shaped phononic crystals to reduce anchor losses of MEMS resonators. In Proceedings of the 2016 IEEE International Ultrasonics Symposium (IUS), Tours, France, 18–21 September 2016; pp. 1–3.

25. Siddiqi, M.W.U.; Lee, J.E.-Y. AlN-on-Si MEMS resonator bounded by wide acoustic bandgap two-dimensional phononic crystal anchors. In Proceedings of the 2018 IEEE International Conference on Micro Electro Mechanical Systems, Belfast, UK, 21–25 January 2018; pp. 727–730.

26. Siddiqi, M.W.U.; Lee, J.E.-Y. Effect of mode order, resonator length, curvature, and electrode coverage on enhancing the performance of biconvex resonators. *J. Micromech. Microeng.* **2018**, *28*, 094002. [CrossRef]

27. Binci, L.; Tu, C.; Zhu, H.; Lee, J.E.-Y. Planar ring-shaped phononic crystal anchoring boundaries for enhancing the quality factor of Lamb mode resonators. *Appl. Phys. Lett.* **2016**, *109*, 203501. [CrossRef]

28. Cowen, A.; Hames, G.; Glukh, K.; Hardy, B. *PiezoMUMPs Design Handbook*; CMC Microsystems: Windsor, ON, Canada, 2014.

29. Tu, C.; Lee, J.E.-Y. VHF-band biconvex AlN-on-silicon micromechanical resonators with enhanced quality factor and suppressed spurious modes. *J. Micromech. Microeng.* **2016**, *26*, 065012. [CrossRef]

30. Zuo, C.; Sinha, N.; Spiegel, J.V.; Piazza, G. Multi-frequency pierce oscillator based on piezoelectric AlN contour-mode MEMS resonators. In Proceedings of the 2008 IEEE International Frequency Control Symposium, Honolulu, HI, USA, 19–21 May 2008; pp. 402–407.

31. Wojciechowski, K.E.; Olsson, R.H.; Tuck, M.R.; Roherty-Osmun, E.; Hill, T.A. Single-chip precision oscillators based on multi-frequency, high-Q aluminum nitride MEMS resonators. In Proceedings of the 2009 International Solid-State Sensors, Actuators and Microsystems Conference, Denver, CO, USA, 21–25 June 2009; pp. 2126–2130.

micromachines

MDPI

Article

Electromagnetically Induced Transparency (EIT) Like Transmission Based on 3 × 3 Cascaded Multimode Interference Resonators

Trung-Thanh Le

International School (VNU-IS), Vietnam National University (VNU), Hanoi 1000, Vietnam; thanh.le@vnu.edu.vn;
Tel.: +84-985-848-193

Received: 1 July 2018; Accepted: 1 August 2018; Published: 9 August 2018

Abstract: We propose a method for generating the electromagnetically induced transparency (EIT) like-transmission by using microring resonator based on cascaded 3 × 3 multimode interference (MMI) structures. Based on the Fano resonance unit created from a 3 × 3 MMI coupler with a feedback waveguide, two schemes of two coupled Fano resonator unit (FRU) are investigated to generate the EIT like transmission. The theoretical and numerical analysis based on the coupled mode theory and transfer matrix is used for the designs. Our proposed structure has advantages of compactness and ease of fabrication. We use silicon waveguide for the design of the whole device so it is compatible with the existing Complementary Metal-Oxide-Semiconductor (CMOS) circuitry foundry. The fabrication tolerance and design parameters are also investigated in this study.

Keywords: optical microring resonator; electromagnetically induced transparency (EIT); multimode interference (MMI); transfer matrix method (TMM); finite difference time difference (FDTD); beam propagation method (BPM)

---

1. Introduction

The electromagnetically induced transparency (EIT) effect is a nonlinear effect found in the interaction process between light and material. The EIT effect has been intensively investigated in recent years [1,2]. The EIT has wide applications such as in quantum information [3], lasing without inversion [4], optical delay, slow light [5], nonlinearity enhancement [6] and precise spectroscopy [7], pushing frontiers in quantum mechanics and photonics and sensing technology [8]. In order to create the EIT effect, there are some suggested approaches.

There is a significant benefit to determine the optical EIT like transmissions with high modulation depth, which is defined by the difference in intensities between the peak and the dip at resonant wavelengths. The EIT was first observed in atomic media [2]. Then, the EIT-like effects were found in optical coupled resonant systems [9–11], mechanics, electrical circuits [12], plasmonics, metamaterials [7,13] and hybrid configurations [14]. In the coupled resonant systems, the basic underlying physical principle is the interference of fields instead of the probability of amplitudes, as in a three-level atomic system [15,16]. Most of the proposed structures so far for the optical EIT generation use metal-insulator-metal (MIM) plasmonic waveguide resonators [17,18], array of fiber optic resonators [19], microspheres [20], metallic arrays of asymmetric dual stripes [21], heptamer-hole array [22], plasmonic nanoring pentamers [23] and microtoroid resonator coupled system [24]. For these systems, the fiber beam splitters, directional couplers or MIM plasmonic waveguide must be used. As a result, such structures bring large size, the complexity of the fabrication process to control exactly the coupling ratios of the directional couplers and sensitivity to fabrication tolerance.

The transparency window of the EIT is caused by reduced absorption, due to the quantum destructive interference between the transitions from the two dressed states, into a common energy level. Similarly, the EIT-like effect generated by optical resonators work by the means of coherent interference between the resonating modes which produce optical transparency inside the absorption window [25]. Compared to the EIT in atomic systems, the analogue of electromagnetically induced transparency with optical resonators based on directional couplers has many remarkable advantages such as simpler structure, smaller device size and easier design. However, due to the small size of these structures, it is challenging to detune optical resonator for controlling the resonant interaction between the two optical pathways and controlling the coupling ratio of the directional couplers [26].

In the literature, only 2×2 directional coupler was used for microring resonator based on the EIT effects [25]. However, such structure is very sensitive to the fabrication. It has a large size and requires a complicated fabrication process. It was shown that the integration of multimode interference (MMI) and resonators can provide new physical characteristics. By using the MMIs, we can overcome the disadvantages of devices based on directional couplers such as compactness, ease of fabrication and large fabrication tolerance [27]. One of such structures is a 3×3 MMI based microring resonator. We have proposed for the first time microring resonator structures based on 3×3 and 4×4 MMI couplers for Fano resonance generation [28–30]. In this study, we further develop new structures based on only cascaded 3×3 multimode interference coupler based microring resonators to produce the EIT resonance like transmissions. The proposed device is analyzed and optimized using the transfer matrix method, the beam propagation method (BPM) and finite difference time difference (FDTD) [31]. A description of the theory behind the use of multimode structures to achieve the FRU and EIT effect is presented in Sections 2 and 3. A brief summary of the results of this research is given in Section 4.

2. Single Fano Resonance Unit (FRU)

Fano resonance can be created by many approaches such as integrated waveguide-coupled microcavities [32,33], prism-coupled square micro-pillar resonators, multimode tapered fiber coupled micro-spheres and Mach Zehnder interferometer (MZI) coupled micro-cavities [34], plasmonic waveguide structure [35,36]. We have proposed integrated photonic circuits for realizing Fano resonance based on 3×3 MMI and 4×4 MMI microring resonator [29,37]. Figure 1a shows a scheme for Fano resonance unit (FRU) based on only one 3×3 MMI coupler with a feedback waveguide. Figure 1b,c shows the FDTD simulation for the FRU with input signal presented at input ports 1 and 2, respectively.

In the time domain, the Fano resonance system created by 3×3 MMI coupler based microring resonator in Figure 1 can be expressed by the coupled mode equations [38]

$$\frac{da_n}{dt} = [j(\omega_0 + \delta\omega_n) - \frac{1}{\tau}]a_n + df_n + dg_{n+1} \tag{1}$$

$$g_n = \exp(j\phi)f_n + da_n \tag{2}$$

where $n = 1, 2$ and $d = j\exp(j\phi/2)/\sqrt{\tau}$; ϕ is the phase of the resonator depending on the feedback waveguide, $\delta\omega_n$ is the nonlinear phase shift, a_n is the amplitude in the resonator mode; f_n, g_n are the complex amplitudes at input and output ports; ω_0 and τ are resonant frequency and lifetime of the resonator.

In the frequency domain, the 3×3 MMI coupler can be described by a transfer matrix $\mathbf{M} = [m_{ij}]_{3\times3}$ which describes the relationships between the input and output complex amplitudes (fields) of the coupler [39]. The length of the MMI coupler is to be $L_{MMI} = L_\pi$, L_π is the beat length of the MMI coupler. The relationship between the output complex amplitudes $b_j (j = 1, 2, 3)$ and the input complex amplitudes $a_i (i = 1, 2, 3)$ of the coupler can be expressed by [39]

$$\begin{pmatrix} b_1 \\ b_2 \\ b_3 \end{pmatrix} = \frac{1}{\sqrt{3}} \begin{pmatrix} -e^{-j2\pi/3} & e^{-j2\pi/3} & -1 \\ e^{-j2\pi/3} & -1 & e^{-j2\pi/3} \\ -1 & e^{-j2\pi/3} & -e^{-j2\pi/3} \end{pmatrix} \begin{pmatrix} a_1 \\ a_2 \\ a_3 \end{pmatrix} = \mathbf{M} \begin{pmatrix} a_1 \\ a_2 \\ a_3 \end{pmatrix} \tag{3}$$

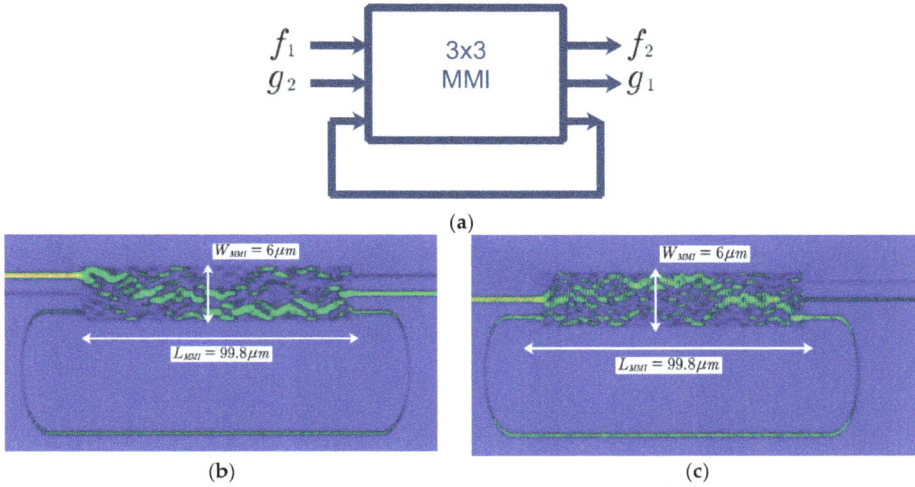

Figure 1. (a) Fano resonance unit created by 3 × 3 MMI (multimode interference) based resonator (b) FDTD simulation for input 1 and (c) finite difference time difference (FDTD) simulation for input 2.

The complex amplitudes at output ports 1 and 2 of the first microring resonator of Figure 1 are given by

$$b_1 = (m_{11} + \frac{m_{13}m_{31}a\exp(jq)}{1 - m_{33}a\exp(jq)})a_1 + (m_{12} + \frac{m_{13}m_{32}a\exp(jq)}{1 - m_{33}a\exp(jq)})a_2 \tag{4}$$

$$b_2 = (m_{21} + \frac{m_{23}m_{31}a\exp(jq)}{1 - m_{33}a\exp(jq)})a_1 + (m_{22} + \frac{m_{23}m_{32}a\exp(jq)}{1 - m_{33}a\exp(jq)})a_2 \tag{5}$$

where $\alpha = \exp(-\alpha_0 L)$ is the transmission loss along the ring waveguide, L is the length of the feedback waveguide and α_0 (dB/cm) is the loss coefficient in the core of the optical waveguide; $\theta = \beta_0 L$ is the phase accumulated over the racetrack waveguide, where $\beta_0 = 2\pi n_{eff}/\lambda$ and n_{eff} is the effective refractive index, λ is the wavelength.

In this study, we use silicon waveguide for the design, where SiO_2 ($n_{SiO_2} = 1.46$) is used as the upper cladding material. The parameters used in the designs are as follows: the waveguide has a standard silicon thickness of $h_{co} = 220$ nm and access waveguide widths are $W_a = 500$ nm for single mode operation. It is assumed that the designs are for the transverse electric (TE) polarization at a central optical wavelength $\lambda = 1550$ nm. In this study, we use the three dimensional beam propagation method (3D-BPM) and Finite Difference Time Domain (FDTD) to design the whole structure [40].

Firstly, we optimize the position of the access waveguide ports of the 3 × 3 MMI coupler to determine the proper matrix of the 3 × 3 MMI coupler expressed by Equation (3). The normalized output powers at output ports of the 3 × 3 MMI varying with the location of input port 1 are shown in Figure 2a. Figure 2b shows the normalized output powers at output ports for different locations of input port 2. Here, the width and length of the MMI coupler are optimized by the BPM simulations to be $W_{MMI} = 6$ μm and $L_{MMI} = 99.8$ μm. As a result, the optimal positions of the input ports 1 and 3 are $p_{1,3} = \mp 2.05$ μm, respectively.

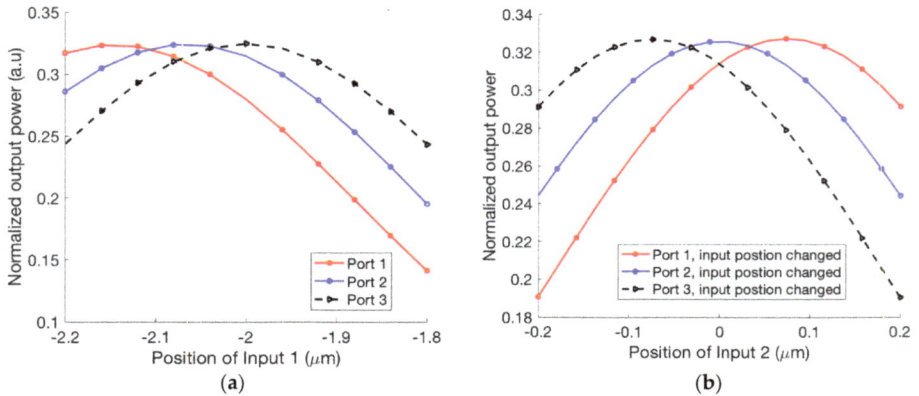

Figure 2. Normalized output powers for different positions of (**a**) input port 1 and (**b**) input port 2.

The phase sensitivity of the output signals to the length variation of the 3×3 MMI coupler based microring resonators is particularly important to device performance. We use the BPM to investigate the effect of the MMI length on the phase sensitivity. Figure 3a shows the phases at output ports of the 3×3 MMI coupler at different MMI lengths. We see that a change of ± 10 nm in MMI length causes a change of $4.7 \times 10^{-4} \pi$ (rad) in output phases. For the existing CMOS circuitry with a fabrication error of ± 5 nm [41], this is feasible and has a very large fabrication tolerance. Similarly, we consider the effect of the positions of input waveguides on the phase sensitivity as shown in Figure 3b. For a fabrication tolerance in the MMI length of ± 50 nm, the fluctuation in phases is nearly unchanged.

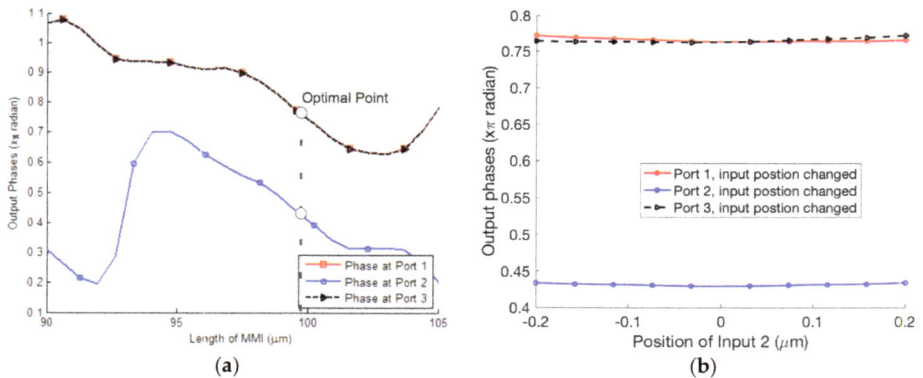

Figure 3. The sensitivity of phases in output ports to (**a**) MMI (multimode interference) length and (**b**) position of input 2.

3. Coupled Fano Resonances and Generation of the EIT Effect

The schemes for coupled Fano resonances to generate the EIT effect is modeled in Figure 4, where single Fano resonance 1 and Fano resonance 2 of Figure 4a is exactly the same and Fano resonance 1 and Fano resonance 2 of Figure 4b is different with an exchange of input ports. We show that by cascading two Fano resonances as shown in Figure 4, the EIT effects can be created. The exchange unit can be realized by using only one 2×2 MMI coupler as shown in reference [42].

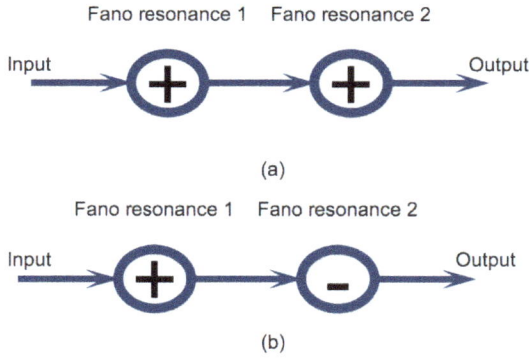

Figure 4. Schemes of Coupled Fano resonances (**a**) bar connect and (**b**) cross connect.

In our case, the Fano resonance 1 and 2 are the FRU created by 3×3 MMI coupler based microring resonator as shown in Figure 1. The first schemes of Figure 4a can be made as shown in Figure 5a and the second scheme of Figure 4b can be made as shown in Figure 5b.

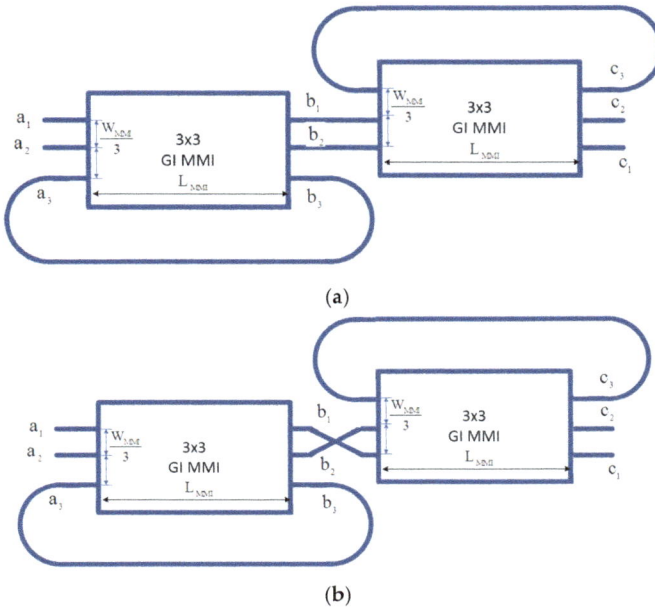

Figure 5. Coupled Fano resonances based on microring resonators based on 3×3 cascaded MMI couplers (**a**) cross without cross-connect and (**b**) bar with cross-connect made from 2×2 MMI coupler [42].

By using analytical analysis, the transmissions at the output ports of Figure 5a, for the input signal presented at input port 1 ($a_2 = 0$) are expressed by

$$T_1 = \left| (m_{11} + \frac{m_{13}m_{31}\alpha e^{j\theta}}{1 - m_{33}\alpha e^{j\theta}})(m_{11} + \frac{m_{13}m_{31}\alpha e^{j\theta}}{1 - m_{33}\alpha e^{j\theta}}) + (m_{12} + \frac{m_{13}m_{32}\alpha e^{j\theta}}{1 - m_{33}\alpha e^{j\theta}})(m_{21} + \frac{m_{23}m_{31}\alpha e^{j\theta}}{1 - m_{33}\alpha e^{j\theta}}) \right|^2 \quad (6)$$

$$T_2 = \left| (m_{21} + \frac{m_{23}m_{31}\alpha e^{j\theta}}{1 - m_{33}\alpha e^{j\theta}})(m_{11} + \frac{m_{13}m_{31}\alpha e^{j\theta}}{1 - m_{33}\alpha e^{j\theta}}) + (m_{22} + \frac{m_{23}m_{32}\alpha e^{j\theta}}{1 - m_{33}\alpha e^{j\theta}})(m_{21} + \frac{m_{23}m_{31}\alpha e^{j\theta}}{1 - m_{33}\alpha e^{j\theta}}) \right|^2 \quad (7)$$

The transmissions at these output ports of Figure 5b, for the input signal presented at input port 1 ($a_2 = 0$) are

$$T_1' = \left| (m_{11} + \frac{m_{13}m_{31}\alpha e^{j\theta}}{1 - m_{33}\alpha e^{j\theta}})(m_{22} + \frac{m_{23}m_{32}\alpha e^{j\theta}}{1 - m_{33}\alpha e^{j\theta}}) + (m_{12} + \frac{m_{13}m_{32}\alpha e^{j\theta}}{1 - m_{33}\alpha e^{j\theta}})(m_{12} + \frac{m_{13}m_{32}\alpha e^{j\theta}}{1 - m_{33}\alpha e^{j\theta}}) \right|^2 \quad (8)$$

$$T_2' = \left| (m_{21} + \frac{m_{23}m_{31}\alpha e^{j\theta}}{1 - m_{33}\alpha e^{j\theta}})(m_{22} + \frac{m_{23}m_{32}\alpha e^{j\theta}}{1 - m_{33}\alpha e^{j\theta}}) + (m_{22} + \frac{m_{23}m_{32}\alpha e^{j\theta}}{1 - m_{33}\alpha e^{j\theta}})(m_{12} + \frac{m_{13}m_{32}\alpha e^{j\theta}}{1 - m_{33}\alpha e^{j\theta}}) \right|^2 \quad (9)$$

For our design, the silicon waveguide is used. The effective refractive index calculated by the FDM (Finite Difference Method) is to be $n_{eff} = 2.416299$ for the TE polarization. It assumed that the loss coefficient of the silicon waveguide is $\alpha = 0.98$ [43], the length of the feedback waveguide is $L_R = 700$ μm [25]. For the first scheme of Figure 5a, the EIT effects shown in Figure 6a can be generated at output ports 1 and 2 while the input signal is at the input port 1. Figure 6b shows the EIT effects are also created at output ports 1 and 2 while the input signal is presented at input port 2. We see that the modulation depth of 80% for these EIT like transmissions have been achieved. As a result, our structure can generate both the W-shape and M-shape transmissions. Such shapes can be useful for optical switching, fast and slow light and sensing applications.

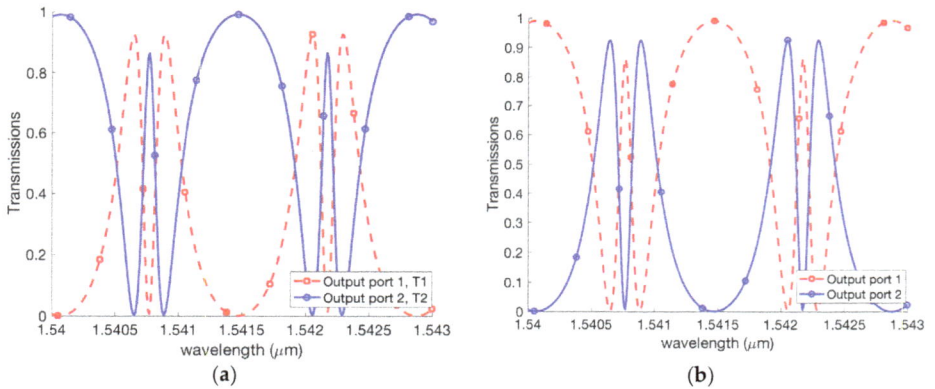

Figure 6. Transmissions of the coupled Fano resonances for Figure 5a with input signal is presented at (**a**) input port 1 and (**b**) input port 2.

For the second scheme of Figure 5b, the EIT effects shown in Figure 7a can be generated at output port 1 and port 2 while input signal is at input port 1. Figure 7b shows the EIT effects are also created at output ports 1 and 2 while the input signal is presented at input port 2.

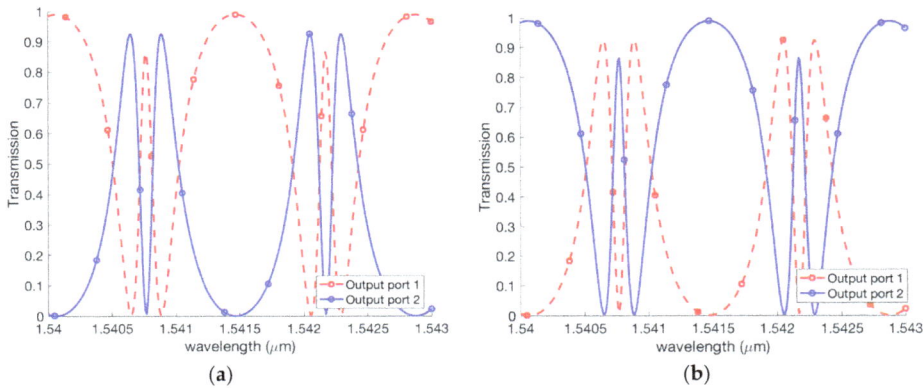

Figure 7. Transmissions of the coupled Fano resonances for Figure 5b with input signal is presented at (**a**) input port 1 and (**b**) input port 2.

In order to verify our proposed analytical theory, we use the FDTD for accurate predictions of the device's working principle. Figure 8 shows the FDTD simulations for the device of Figure 5a,b for input signal at port 1, respectively. Figure 9 shows the FDTD of Figure 5a,b for input signal at port 2. In our FDTD simulations, we take into account of the refractive index of silicon material calculated by using the Sellmeier equation [44,45]:

$$n^2(\lambda) = \varepsilon + \frac{A}{\lambda^2} + \frac{B\lambda_1^2}{\lambda^2 - \lambda_1^2} \tag{10}$$

where $\varepsilon = 11.6858$, $A = 0.939816$ mm^2, $B = 8.10461 \times 10^{-3}$ and $\lambda_1 = 1.1071$ mm.

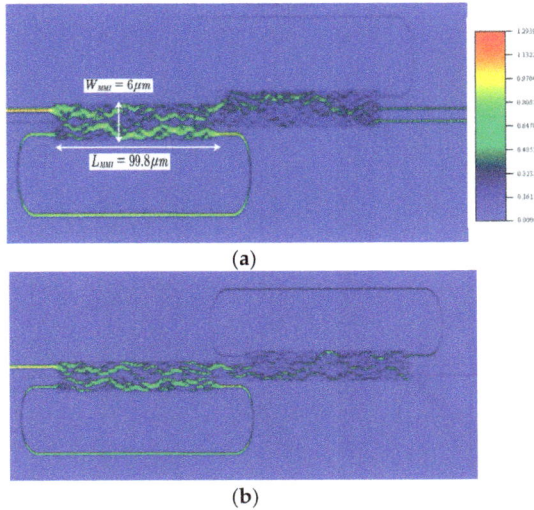

Figure 8. FDTD simulations for input signal at input port 1 for the EIT scheme of Figure 5a,b at wavelength $\lambda = 1550$ nm.

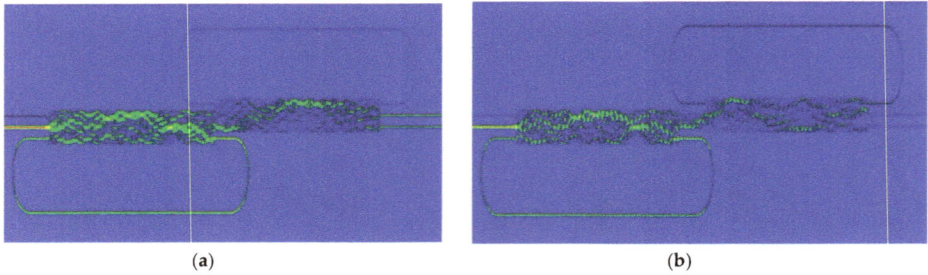

(a)   (b)

Figure 9. FDTD simulations for input signal at input port 2 for the EIT scheme of Figure 5a,b $\lambda = 1550$ nm.

In our FDTD simulations, a Gaussian light pulse of 15 fs pulse width is launched from the input to investigate the transmission characteristics of the device. The grid sizes $\Delta x = \Delta y = 5$ nm and $\Delta z = 10$ nm are chosen [46].

For the purpose of comparing the theoretical and FDTD analysis, we investigate a comparison of the EIT like transmission effect between the theory and FDTD simulations. It is shown that the FDTD simulation has a good agreement with our theoretical analysis as presented in Figure 10.

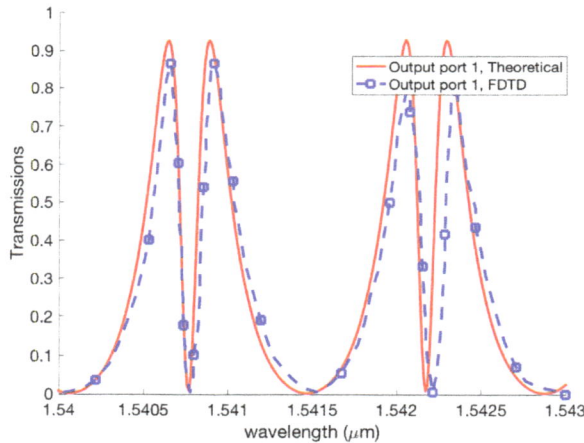

Figure 10. Comparison of theoretical and FDTD simulations.

4. Conclusions

We have presented a new method for the generation of the EIT effect based on coupled 3×3 MMI based microring resonators. Both of the M-shape and W-shape like transmissions are created. The device based on silicon waveguide, that is compatible with the existing CMOS circuitry, has been optimally designed. Our FDTD simulations show a good agreement with the theoretical analysis based on the transfer matrix method. The EIT effect can be determined based on these structures with advantages of ease of fabrication and large fabrication tolerance.

Funding: This research is funded by Ministry of Natural Resources and Environment of Vietnam under the project BĐKH.30/16-20.

Conflicts of Interest: The author declares no conflict of interest.

References

1. Zhou, X.; Zhang, L.; Pang, W.; Zhang, H.; Yang, Q.; Zhang, D. Phase characteristics of an electromagnetically induced transparency analogue in coupled resonant systems. *New J. Phys.* **2013**, *15*, 103033. [CrossRef]
2. Fleischhauer, M.; Imamoglu, A.; Marangos, J.P. Electromagnetically induced transparency: Optics in coherent media. *Rev. Mod. Phys.* **2005**, *77*, 633–673. [CrossRef]
3. Lvovsky, A.I.; Sanders, B.C.; Tittel, W. Optical quantum memory. *Nat. Photonics* **2009**, *3*, 706–714. [CrossRef]
4. Harris, S.E. Lasers without inversion: Interference of lifetime-broadened resonances. *Phys. Rev. Lett.* **1989**, *62*, 1033–1036. [CrossRef] [PubMed]
5. Chin, S.; Thévenaz, L. Tunable photonic delay lines in optical fibers. *Laser Photonics Rev.* **2012**, *6*, 724–738. [CrossRef]
6. Tanji-Suzuki, H.; Chen, W.; Landig, R.; Simon, J.; Vuletić, V. Vacuum-induced transparency. *Science* **2011**, *361*. [CrossRef]
7. Tassin, P.; Zhang, L.; Zhao, R.; Jain, A.; Koschny, T.; Soukoulis, C.M. Electromagnetically induced transparency and absorption in metamaterials: The radiating two-oscillator model and its experimental confirmation. *Phys. Rev. Lett.* **2012**, *109*, 187401. [CrossRef] [PubMed]
8. Wu, D.; Liu, Y.; Yu, L.; Yu, Z.; Chen, L.; Li, R.; Ma, R.; Liu, C.; Zhang, J.; Ye, H. Plasmonic metamaterial for electromagnetically induced transparency analogue and ultra-high figure of merit sensor. *Sci. Rep.* **2017**, *7*, 45210. [CrossRef] [PubMed]
9. Totsuka, K.; Kobayashi, N.; Tomita, M. Slow light in coupled-resonator-induced transparency. *Phys. Rev. Lett.* **2007**, *98*, 213904. [CrossRef] [PubMed]
10. Chremmos, I.; Schwelb, O. *Photonic Microresonator Research and Applications*; Springer: New York, NY, USA, 2010.
11. Xu, Q.; Sandhu, S.; Povinelli, M.L.; Shakya, J.; Fan, S.; Lipson, M. Experimental realization of an on-chip all-optical analogue to electromagnetically induced transparency. *Phys. Rev. Lett.* **2006**, *96*, 123901. [CrossRef] [PubMed]
12. Garrido Alzar, C.L.; Martinez, M.A.G.; Nussenzveig, P. Classical analog of electromagnetically induced transparency. *Am. J. Phys.* **2002**, *70*, 37–41. [CrossRef]
13. Liu, N.; Langguth, L.; Weiss, T.; Kästel, J.; Fleischhauer, M.; Pfau, T.; Giessen, H. Plasmonic analogue of electromagnetically induced transparency at the drude damping limit. *Nat. Mater.* **2009**, *8*, 758–762. [CrossRef] [PubMed]
14. Weis, P.; Garcia-Pomar, J.L.; Beigang, R.; Rahm, M. Hybridization induced transparency in composites of metamaterials and atomic media. *Opt. Express* **2011**, *19*, 23573–23580. [CrossRef] [PubMed]
15. Smith, D.D.; Chang, H.; Fuller, K.A.; Rosenberger, A.T.; Boyd, R.W. Coupled-resonator-induced transparency. *Phys. Rev. A* **2004**, *69*, 063804. [CrossRef]
16. Peng, B.; Özdemir, Ş.K.; Chen, W.; Nori, F.; Yang, L. What is and what is not electromagnetically induced transparency in whispering-gallery microcavities. *Nat. Commun.* **2014**, *5*, 5082. [CrossRef] [PubMed]
17. Chen, Z.; Song, X.; Duan, G.; Wang, L.; Yu, L. Multiple fano resonances control in mim side-coupled cavities systems. *IEEE Photonics J.* **2015**, *7*, 1–5. [CrossRef]
18. Wang, Y.; Li, S.; Zhang, Y.; Yu, L. Independently formed multiple fano resonances for ultra-high sensitivity plasmonic nanosensor. *Plasmonics* **2018**, *13*, 107–113. [CrossRef]
19. Li, J.; Qu, Y.; Wu, Y. Add-drop double bus microresonator array local oscillators for sharp multiple fano resonance engineering. *J. Appl. Phys.* **2018**, *123*, 104305. [CrossRef]
20. Dong, C.-H.; Zou, C.-L.; Xiao, Y.-F.; Cui, J.-M.; Han, Z.-F.; Guo, G.-C. Modified transmission spectrum induced by two-mode interference in a single silica microsphere. *J. Phys. B At. Mol. Opt. Phys.* **2009**, *42*, 215401. [CrossRef]
21. Chen, Y.-T.; Chern, R.-L.; Lin, H.-Y. Multiple fano resonances in metallic arrays of asymmetric dual stripes. *Appl. Opt.* **2010**, *49*, 2819–2826. [CrossRef] [PubMed]
22. He, J.; Ding, P.; Wang, J.; Fan, C.; Liang, E. Double fano-type resonances in heptamer-hole array transmission spectra with high refractive index sensing. *J. Mod. Opt.* **2015**, *62*, 1241–1247. [CrossRef]
23. Liu, H.; Leong Eunice Sok, P.; Wang, Z.; Si, G.; Zheng, L.; Liu, Y.J.; Soci, C. Multiple and multipolar fano resonances in plasmonic nanoring pentamers. *Adv. Opt. Mater.* **2013**, *1*, 978–983. [CrossRef]
24. Liang, W.; Yang, L.; Poon, J.K.; Huang, Y.; Vahala, K.J.; Yariv, A. Transmission characteristics of a fabry-perot etalon-microtoroid resonator coupled system. *Opt. Lett.* **2006**, *31*, 510–512. [CrossRef] [PubMed]

25. Liu, Y.-C.; Li, B.-B.; Xiao, Y.-F. Electromagnetically induced transparency in optical microcavities. *Nanophotonics* **2017**, *6*, 789–811. [CrossRef]

26. Le, T.T.; Cahill, L.W.; Elton, D. The design of 2 × 2 SOI MMI couplers with arbitrary power coupling ratios. *Electron. Lett.* **2009**, *45*, 1118–1119. [CrossRef]

27. Soldano, L.B.; Pennings, E.C.M. Optical multi-mode interference devices based on self-imaging: Principles and applications. *IEEE J. Lightwave Tech.* **1995**, *13*, 615–627. [CrossRef]

28. Le, T.-T.; Cahill, L. Generation of two fano resonances using 4 × 4 multimode interference structures on silicon waveguides. *Opt. Commun.* **2013**, *301*, 100–105. [CrossRef]

29. Le, T.-T. Fano resonance based on 3 × 3 multimode interference structures for fast and slow light applications. *Int. J. Microwave Opt. Technol.* **2017**, *13*, 406–412.

30. Le, D.-T.; Le, T.-T. Fano resonance and EIT-like effect based on 4 × 4 multimode interference structures. *Int. J. Appl. Eng. Res.* **2017**, *12*, 3784–3788.

31. Huang, W.P.; Xu, C.L.; Lui, W.; Yokoyama, K. The perfectly matched layer (PML) boundary condition for the beam propagation method. *IEEE Photonics Technol. Lett.* **1996**, *8*, 649–651. [CrossRef]

32. Fan, S. Sharp asymmetric line shapes in side-coupled waveguide-cavity systems. *Appl. Phys. Lett.* **2002**, *80*, 908–910. [CrossRef]

33. Le, D.-T.; Le, T.-T. Coupled resonator induced transparency (CRIT) based on interference effect in 4 × 4 MMI coupler. *Int. J. Comput. Syst.* **2017**, *4*, 95–98.

34. Hon, K.Y.; Poon, A. Silica polygonal micropillar resonators: Fano line shapes tuning by using a mach-zehnder interferometer. In Proceedings of the Lasers and Applications in Science and Engineering, San Jose, CA, USA, 23 February 2006.

35. Chen, Z.-Q.; Qi, J.-W.; Jing, C.; Li, Y.-D. Fano resonance based on multimode interference in symmetric plasmonic structures and its applications in plasmonic nanosensors. *Chin. Phys. Lett.* **2013**, *30*, 057301. [CrossRef]

36. Zhang, B.-H.; Wang, L.-L.; Li, H.-J.; Zhai, X.; Xia, S.-X. Two kinds of double fano resonances induced by an asymmetric MIM waveguide structure. *J. Opt.* **2016**, *18*, 065001. [CrossRef]

37. Le, D.-T.; Do, T.-D.; Nguyen, V.-K.; Nguyen, A.-T.; Le, T.-T. Sharp asymmetric resonance based on 4 × 4 multimode interference coupler. *Int. J. Appl. Eng. Res.* **2017**, *12*, 2239–2242.

38. Fan, S.; Suh, W.; Joannopoulos, J.D. Temporal coupled-mode theory for the fano resonance in optical resonators. *J. Opt. Soc. Am. A* **2003**, *20*, 569–572. [CrossRef]

39. Le, T.-T.; Cahill, L. Microresonators based on 3 × 3 restricted interference MMI couplers on an soi platform. In Proceedings of the Annual Meeting Conference (LEOS 2009), Belek-Antalya, Turkey, 4–8 October 2009.

40. Taflove, A.; Hagness, S.C. *Computational Electrodynamics: The Finite-Difference Time-Domain Method*, 3rd ed.; Artech House: Norwood, MA, USA, 2005.

41. Bogaerts, W.; Heyn, P.D.; Vaerenbergh, T.V. Silicon microring resonators. *Laser Photonics Rev.* **2012**, *6*, 47–73. [CrossRef]

42. Le, T.-T. The design of optical signal transforms based on planar waveguides on a silicon on insulator platform. *Int. J. Eng. Technol.* **2010**, *2*, 245–251.

43. Xia, F.; Sekaric, L.; Vlasov, Y.A. Mode conversion losses in silicon-on-insulator photonic wire based racetrack resonators. *Opt. Express* **2006**, *14*, 3872–3886. [CrossRef] [PubMed]

44. Palik, E.D. *Handbook of Optical Constants of Solids*; Academic Press: San Diego, CA, USA, 1998.

45. Le, T.-T. Highly sensitive sensor based on 4 × 4 multimode interference coupler with microring resonators. *J. Opt. Adv. Mater.* **2018**, *5*, 264–270.

46. Le, D.-T.; Nguyen, M.-C.; Le, T.-T. Fast and slow light enhancement using cascaded microring resonators with the sagnac reflector. *Optik Int. J. Light Elect. Opt.* **2017**, *131*, 292–301. [CrossRef]

micromachines

MDPI

Article

Utilization of 2:1 Internal Resonance in Microsystems

Navid Noori, Atabak Sarrafan, Farid Golnaraghi and Behraad Bahreyni *

School of Mechatronic Systems Engineering, Simon Fraser University, Surrey, BC V5A 1S6, Canada; nnoori@sfu.ca (N.N.); asarrafa@sfu.ca (A.S.); mfgolnar@sfu.ca (F.G.)
* Correspondence: behraad@ieee.org; Tel.: +1-778-782-8694

Received: 3 August 2018; Accepted: 6 September 2018; Published: 8 September 2018

Abstract: In this paper, the nonlinear mode coupling at 2:1 internal resonance has been studied both analytically and experimentally. A modified micro T-beam structure is proposed, and the equations of motion are developed using Lagrange's energy method. A two-variable expansion perturbation method is used to describe the nonlinear behavior of the system. It is shown that in a microresonator with 2:1 internal resonance, the low-frequency mode is autoparametrically excited after the excitation amplitude reaches a certain threshold. The effect of damping on the performance of the system is also investigated.

Keywords: 2:1 internal resonance; energy transfer; micromachined resonators; nonlinear modal interactions; perturbation method

1. Introduction

Microresonators are microfabricated devices that can be operated at their resonance. These microdevices are used in a variety of applications including timing references, filters, sensors, actuators, etc. [1]. Microresonators are typically used within their linear range of operation. Nevertheless, there has been an increasing interest in operating the microresonators at the nonlinear mode of operation to enhance their performance [2–5]. Unavoidable nonlinearities can be found in many micro or macro systems. They can cause severe impact on performance of high quality-factor (Q) MEMS devices. In some cases, neglecting the presence of nonlinearities can lead to erroneous predictions of system's dynamic [6]. It is not always the case that engineers avoid the nonlinearities because of the degradation in the system's performance and the unwanted outcomes. MEMS designers can also beneficially exploit nonlinearities, e.g., mechanical or electrical nonlinearities, in the design of microdevices for various purposes, e.g., sensing, actuation, timing, and signal processing [7].

Nonlinear mode coupling is one of the outcomes of the presence of nonlinearities in the system. Nonlinear mode coupling results in transfer of energy from an intentionally excited mode to other modes of vibration. One of the mechanisms of nonlinear mode coupling is internal resonance. Internal resonance (also known as autoparametric excitation) refers to the transfer of energy from one vibrational mode to another mode where the resonance frequencies of these vibrational modes are commensurable or nearly commensurable. Internal resonance can be used in various applications including mass sensing, inertial sensing [8], energy harvesting [9], and noise suppression [10].

In a system with nonlinear 2:1 internal resonance: (1) there exists a frequency ratio of 2:1 between the natural frequencies of two resonant modes, and (2) quadratic nonlinearities couple the vibrational modes [11]. Due to this nonlinear mode coupling, energy can be channelled from one mode to another mode. There is an interesting nonlinear phenomenon in the systems with 2:1 internal resonance, which is known as saturation. When the system is excited at its primary mode, the amplitude of this mode increases linearly with an increase in the excitation amplitude until the modal amplitude reaches a specific threshold. After this point, the amplitude of the primary mode remains at a constant value and the excessive energy acquired by the increase in the excitation amplitude channels to the secondary mode. One of the simplest examples of a system with 2:1 internal resonance is a spring-pendulum system where energy is being exchanged between spring mode and pendulum mode [12–14]. A similar study to investigate the internal resonance in a micro H-shaped microdevice has been done by Sarrafan et al. [15].

In this paper, a microresonator with 2:1 internal resonance is introduced. The mathematical model of the system is developed and solved by using a perturbation method. Reduced-order analysis of the structure in CoventorWare© software (ver. 2012, Coventor, Inc., Cary, NC, USA) is also explained. Finally, fabricated microdevice is characterized, and the experimental results showing nonlinear mode coupling are thoroughly discussed.

2. Materials and Methods

A modified micro T-beam structure is designed to operate based on the principle of nonlinear 2:1 internal resonance. The schematic of the microdevice is shown in Figure 1. This design idea originated from the T-beam design in [16]. The modified T-beam structure consists of three beams: (1) the drive beam (bottom beam) which is anchored to the substrate and used for actuating the structure, (2) the narrowed beam which is connected to the center of the drive beam and has a relatively low stiffness, and (3) the sense beam which is connected to the narrowed beam and is used for sensing the response of the system.

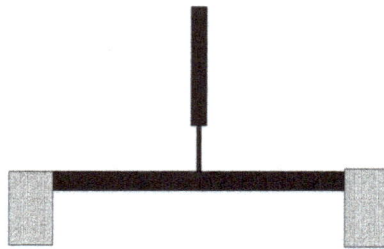

Figure 1. Schematic view of the modified micro T-beam.

Figure 2 shows the desired mode shapes of the structure from finite element method (FEM) simulations in ANSYS® software. The structure is designed to have a frequency ratio of 2:1 between its first and second structural modes. It is expected that exciting the system at its second resonance frequency excites the first vibrational mode autoparametrically.

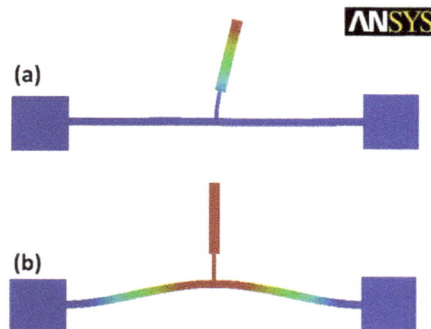

Figure 2. In-plane mode shapes of the micro T-beam structure obtained from ANSYS® FEM simulation: (**a**) the first mode (**b**) the second mode.

To better understand the nonlinear mode coupling and energy transfer between vibrational modes, the equations of motion of the system can be solved by perturbation method. The first step in this process is to model the system mathematically. The modified T-beam structure can be described by using lumped elements, shown in Figure 3. In this model, m_1 and m_2 represent the effective mass of the drive beam and the sense beam, respectively. Similarly, c_i and k_i are the effective damping coefficient and the spring constant of the beams, respectively. By using Lagrange's energy method, the equations of motion can be written as:

$$(m_1 + m_2)\ddot{r}_1 + c_1\dot{r}_1 + k_1 r_1 - m_2 r_2((\ddot{\theta}_1 \sin(\theta_2) + \dot{\theta}_1^2 \cos(\theta_2) + 2\dot{\theta}_1\dot{\theta}_2\cos(\theta_2))$$
$$-m_2 r_2(\ddot{\theta}_2 sin(\theta_2) + \dot{\theta}_2^2 \cos(\theta_2)) - (m_1 + m_2)r_1\dot{\theta}_1^2 = F_{drive}(t)$$

$$(1)$$

$$m_2 r_2^2\ddot{\theta}_2 + c_2\dot{\theta}_2 + k_2\theta_2 - m_2 r_2\ddot{r}_1 \sin(\theta_2) + 2m_2\dot{r}_1 r_2\dot{\theta}_1 \cos(\theta_2)$$
$$+m_2 r_1 r_2(\ddot{\theta}_1 \cos(\theta_2) + \dot{\theta}_1^2 \sin(\theta_2)) = -m_2 r_2^2\ddot{\theta}_1$$

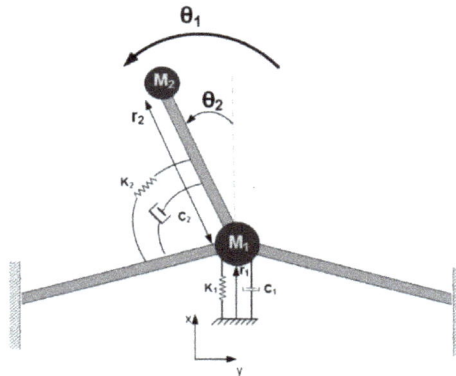

Figure 3. Lumped element representation of micro T-beam structure.

These equations will then be non-dimensionalized and scaled. A perturbation method named as two-variable expansion method is used to study the nonlinear behaviour of the system. Due to the lengthy nature of the perturbation method, only the final solution of the system is provided here, with additional details about each step in the perturbation method provided in [5]. The final perturbation solution of the modified T-beam structure for ρ and θ, the non-dimensionalized amplitudes of the drive beam and the sense beam, respectively, are found from [5]:

$$\rho = \frac{2\omega_2}{\omega_1^2}\sqrt{(\sigma_1 + \sigma_2)^2 + \mu_2^2}\cos(\Omega_1\tau - \gamma_1) + O(\varepsilon)$$

$$(2)$$

$$\theta = \frac{1}{\omega_2}\sqrt{\frac{\Lambda_1 \pm \sqrt{f_1^2\omega_1^6 - \Lambda_2^2}}{m\omega_1}}\cos(\tfrac{1}{2}\Omega_1\tau - \tfrac{\gamma_1 + \gamma_2}{2}) + O(\varepsilon)$$

Parameters in Equation (2) are defined in Table 1.

Figure 4 shows the simulated nonlinear frequency response of the system from perturbation solution. As it can be seen, as the energy starts to transfer between the modes, the amplitude of the drive mode drops and the amplitude in sense mode grows. It can also be seen that a nearly flat region is formed in the frequency response of the sense mode. Figure 5 illustrates the saturation phenomenon. It can be seen that the system behaves linearly before the amplitude of the drive mode reaches a certain point and there is no energy transfer between vibrational modes. However, the system starts to behave nonlinearly as the amplitude in the drive mode reaches the threshold.

Table 1. Definition of nondimensionlized parameters for Equation (2).

Nondimensionalized Parameters	Symbol
Drive mode frequency	ω_1
Sense mode frequency	ω_2
Perturbation parameter	ε
Detuning frequency ($\Omega_1 = \omega_1 + \varepsilon\sigma_1$)	σ_1
Detuning frequency ($\omega_1 = 2\omega_2 + \varepsilon\sigma_2$)	σ_2
Drive beam damping	γ_1
Sense beam damping	γ_2
Excitation force amplitude	f_1
Excitation force frequency	Ω_1
$\omega_2[4\sigma_1(\sigma_1 + \sigma_2) - 2\mu_1\mu_2]$	Λ_1
$\omega_2[2\mu_1(\sigma_1 + \sigma_2) + 4\sigma_1\mu_2]$	Λ_2

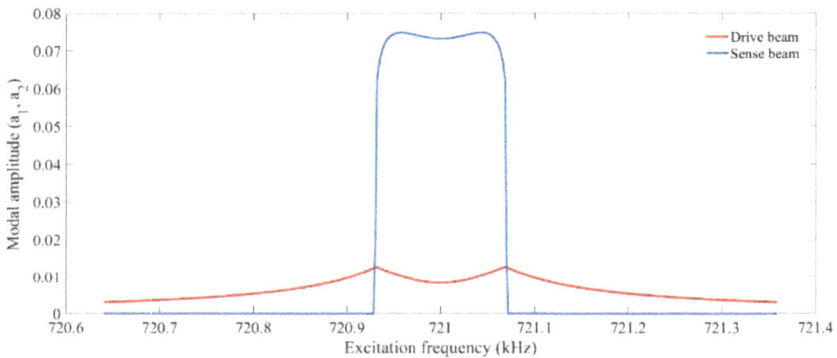

Figure 4. Simulated Nonlinear frequency sweep achieved from two-variable perturbation solution.

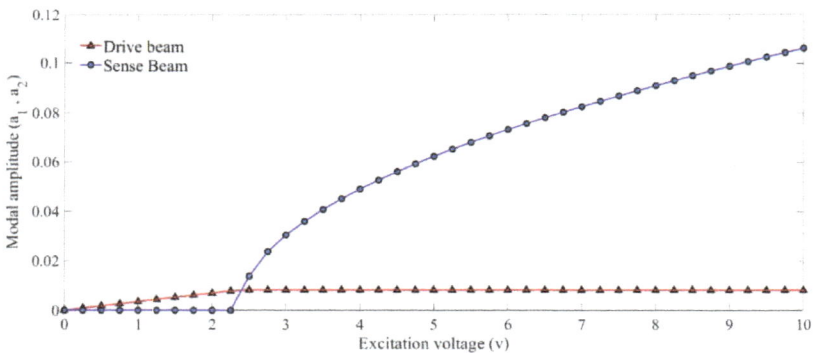

Figure 5. Simulated saturation curve by two-variable perturbation solution.

Numerical simulation in CoventorWare© to model the dynamical behaviour can be helpful for the proper design of the system. FEM analysis is also an essential step in the design process of the system to ensure that resonance frequencies of the two desired structural modes are close to the 2:1 ratio ($\omega_2 = 2\omega_1$). Dimensional adjustments are done based on these results from FEM analysis to ensure the target 2:1 frequency ratio. Architect module in CoventorWare© is being used to perform a reduced-order modelling and simulation of the system. Figure 6 shows the schematic view of the modelled system in the CoventorWare© Architect.

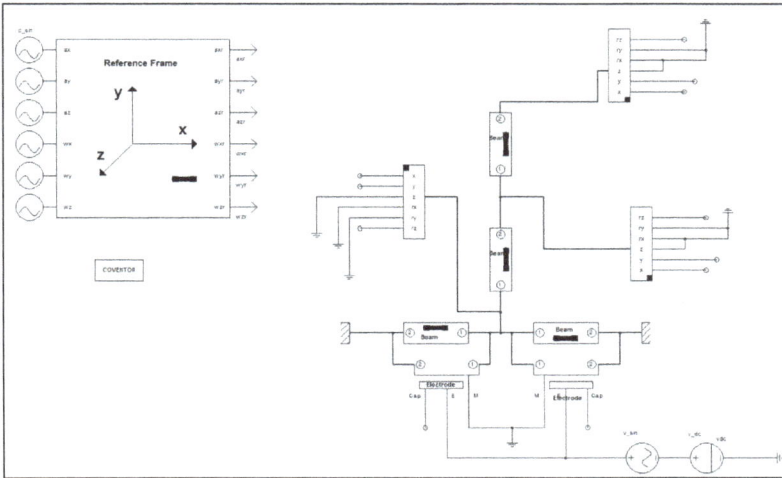

Figure 6. System schematic in CoventorWare© Architect.

The electrostatic excitation is used in the simulations. A 40 V DC voltage is applied to the electrodes, and the drive beam is excited to reach its resonance by applying appropriate AC voltage to the drive electrode. Figure 7 shows the time response of the system with energy transfer between vibrational modes. As it can be seen in the time response-similar to results from perturbation solution, when internal resonance begins to happen, energy starts to exchange between vibrational modes. The drive mode amplitude starts to drop, and the amplitude of the sense mode simultaneously starts to grow exponentially until the system reaches to the steady state. The exponential growth of the sense mode amplitude reveals the absence of damping during the transfer of energy between modes.

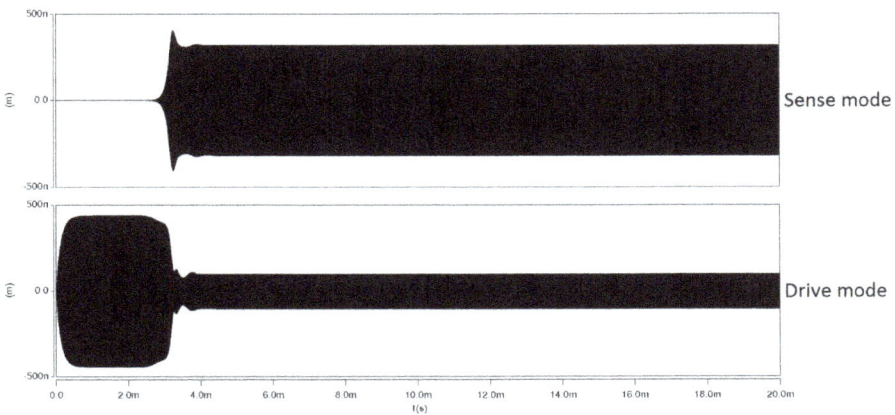

Figure 7. Time-domain response from transient simulation in CoventorWare© showing the transfer of energy between vibrational modes.

It is also expected that by increasing the excitation amplitude, the coupling between vibrational modes becomes stronger. Therefore, the higher amplitudes can be observed in both the drive and sense modes. Figure 8 demonstrates the nonlinear frequency response of the system for different excitation amplitudes.

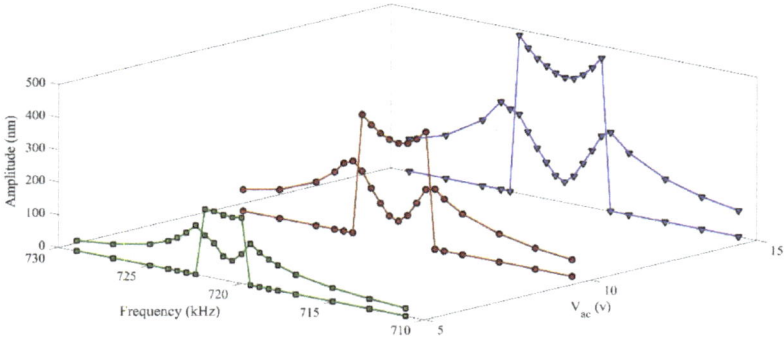

Figure 8. Nonlinear frequency curves in CoventorWare© showing the nonlinear mode coupling between the drive and sense modes.

With the help of the perturbation solution and the CoventorWare© simulations, the modified micro T-beam structure is designed. The microdevice is fabricated by Silicon-on-Insulator Multi-User MEMS Processes (SOIMUMPS). SOIMUMPS is a general purpose microfabrication process introduced by MEMSCAP for micromachining of devices with highly planar surfaces in a SOI framework. This process is a simple 4-mask level SOI patterning and etching. It is a great choice of fabrication for proof of concept purpose. This process has a minimum feature size of 2 μm and the minimum gap between any two silicon parts is also 2 μm. More details about this process can be found in [17]. Figure 9 shows the final fabricated structure.

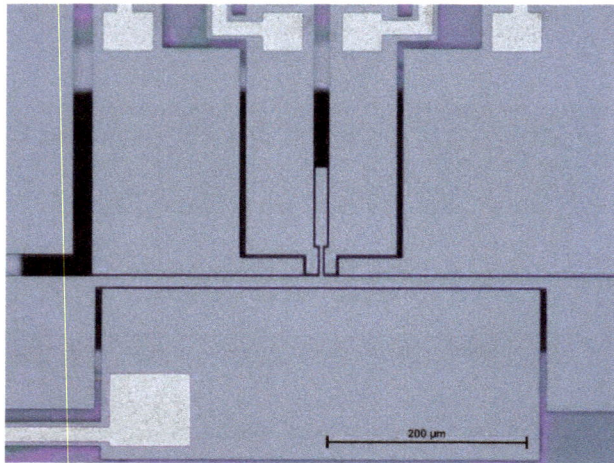

Figure 9. Fabricated device using SOIMUMPS (Silicon-on-Insulator Multi-User MEMS Processes) process.

The next step is to perform experimental tests to investigate the nonlinear response of the fabricated structure. Figure 10 depicts the test setup used for the nonlinear frequency sweep test. The experimental setup used to conduct the frequency sweep tests consists of (1) the fabricated modified micro T-beam, (2) vacuum chamber, (3) a DC voltage source, (4) a function generator for excitation, (5) a signal amplifier, (6) spectrum analyzer to monitor the output signal. As it can be seen, the system is being excited to reach its resonance by exciting the drive beam electrostatically. The response of the system is being sensed through the electrostatic electrodes beside the sense beam and is then being amplified before reaching the spectrum analyzer for monitoring.

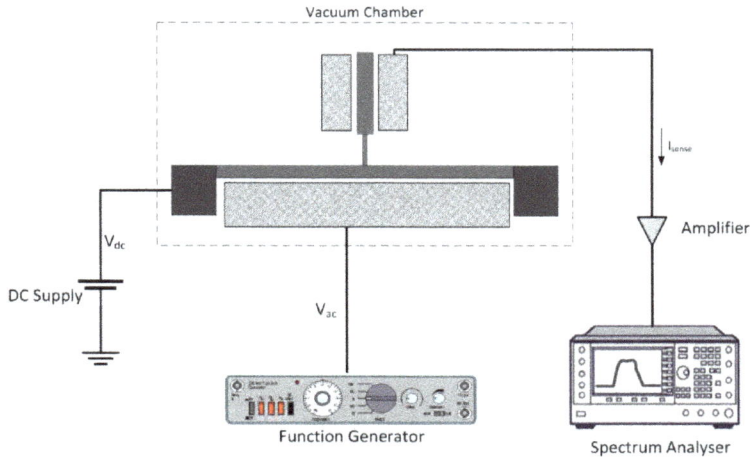

Figure 10. Experimental setup for frequency sweep. Both excitation and sensing are done by electrostatic transduction.

Resonance frequencies of the fabricated structure are also measured by a network analyzer and specified to be 361.135 kHz (first mode) and 722.590 kHz (second mode). These measurements imply a nearly ideal frequency ratio of 2.001. In the next section, the experimental frequency sweeps and also the effect of damping on the performance of the system are discussed.

3. Results and Discussion

The drive beam is actuated by an AC signal with a frequency near its resonance frequency in the range of 714 kHz and 724 kHz. Both forward and backward frequency sweeps are acomplished to investigate the nonlinear performance of the structure. Figure 11 shows the sense beam's response in forward and backward frequency sweeps. It can also be observed in this figure that there is an overlapping region between forward and backward frequency sweeps. This region relates to the frequency range that response of the system is not dependent on the direction of the sweep.

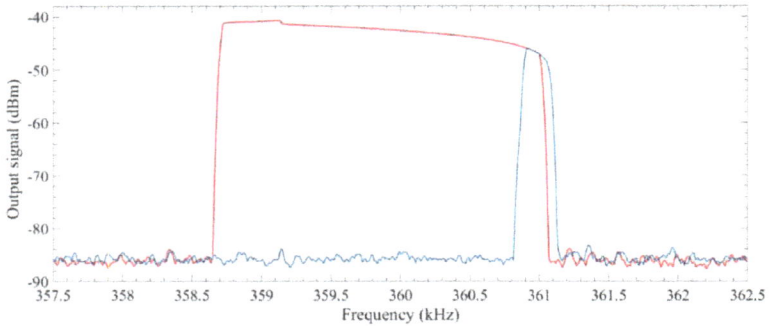

Figure 11. Forward (blue) and backward (red) frequency sweeps showing the nonlinear response of the system.

The next step is to investigate the effect of damping on the nonlinear response of the system. Damping of a resonator can be represented by the quality factor of the system (Q-factor). Quality factor is one of the most important parameters of microresonators which directly relates to the resonance amplitude of the microresonator. In a linear microresonator, a linear relation is expected between Q-factor and amplitude at resonance. Figure 12 shows quality factor of the sense mode of the system within its linear range of operation at different operating pressures. As can be seen, viscous damping dominates the energy loss at pressures above ~100 mTorr. At lower pressures, other sources of energy loss, such as support losses or thermoelastic damping, dominate. As these loss mechanisms are independent of pressure, the quality factor plateaus at pressures less than ~70 mTorr.

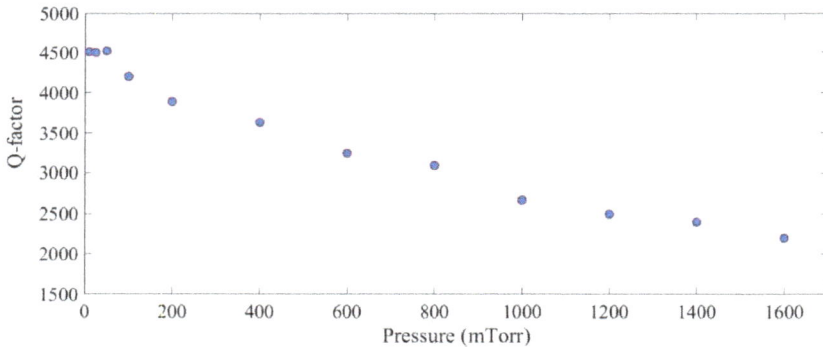

Figure 12. Quality factor of the sense mode of the structure while operating in its linear region.

This figure also shows that the quality factor of the system operating in the linear region varies between 4500 and 2200 in the operating pressure of nearly zero (10 mTorr) to 1.6 Torr. A set of experiment is conducted to investigate the effect of operating pressure on the response of the microresonator in the presence of the 2:1 internal resonance. To conduct this test, the vacuum chamber's pressure is being changed slowly from near vacuum condition (here 10 mTorr) to 1.6 mTorr. After fixing the pressure, frequency sweeps are performed to show the internal resonance phenomenon. Figures 13 and 14 show the nonlinear response of the system in different operating pressures for both backward and forward frequency sweeps.

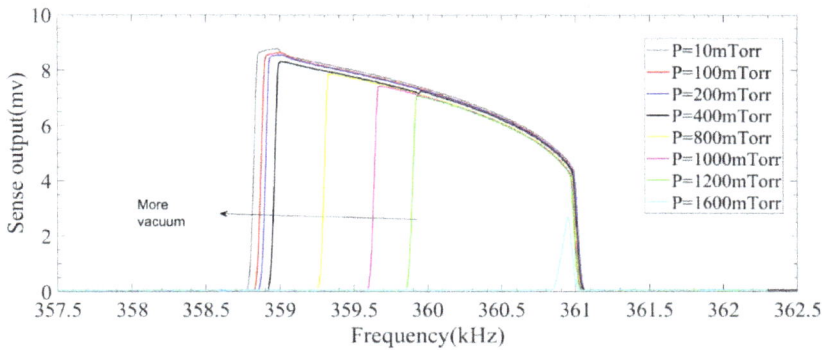

Figure 13. Effect of pressure on the performance of the system in the presence of internal resonance in a backward frequency sweep.

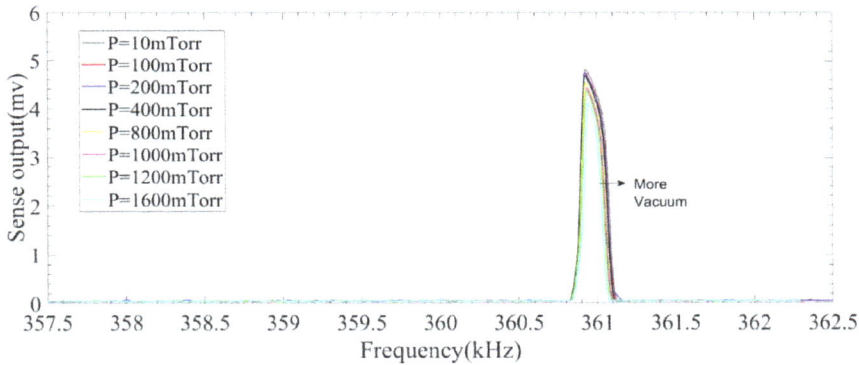

Figure 14. Effect of pressure on the performance of the system in the presence of internal resonance in a forward frequency sweep.

These figures reveal that unlike the linear case, the amplitude of response in the presence of 2:1 internal resonance does not change significantly for pressures below 1600 mTorr. However, the bandwidth of the response becomes smaller as the operating pressure increases. These results show that microresonators working under internal resonance do not need to be necessarily packaged in near vacuum which can significantly reduce the packaging costs of these MEMS devices.

4. Conclusions

A modified micro T-beam with the 2:1 internal resonance was proposed and designed with the help of perturbation solutions and the nonlinear analysis in CoventorWare© Architect. The designed structure was then fabricated by a SOIMUMPS process. Experimental tests on the fabricated structure with a nearly perfect frequency ratio of 2:1 verified the simulation results qualitatively. A significant response enhancement in both forward and backward frequency sweeps were also observed in these results. The response of the system in different operating pressures was studied. It showed that unlike the linear response of the system, response amplitude under the 2:1 internal resonance does not depend significantly on the operating pressure.

Author Contributions: Conceptualization, F.G., B.B.; Methodology, N.N., A.S., F.G.; Investigation, N.N., A.S., B.B.; Writing-Original Draft Preparation, N.N.; Writing-Review & Editing, A.S., B.B., F.G.; Supervision, F.G., B.B.

Funding: This research was funded by the Natural Sciences and Engineering Research Council of Canada, project STPGP 493983.

Acknowledgments: Access to fabrication services and simulation tools was provided through CMC Microsystems. The authors would like to acknowledge the help and guidance provided by Dr Soheil Azimi during the testing of the devices.

Conflicts of Interest: The authors declare no conflicts of interest

References

1. Abdolvand, R.; Bahreyni, B.; Lee, J.E.-Y.; Nabki, F. Micromachined Resonators: A Review. *Micromachines* **2016**, *7*, 160. [CrossRef]
2. Vyas, A.; Peroulis, D.; Bajaj, A.K. A Microresonator Design Based on Nonlinear 1:2 Internal Resonance in Flexural Structural Modes. *J. Microelectromech. Syst.* **2009**, *18*, 744–762. [CrossRef]
3. Antonio, D.; Zanette, D.H.; López, D. Frequency stabilization in nonlinear micromechanical oscillators. *Nat. Commun.* **2012**, *3*, 806. [CrossRef] [PubMed]
4. Fatemi, H.; Shahmohammadi, M.; Abdolvand, R. Ultra-stable nonlinear thin-film piezoelectric-on-substrate oscillators operating at bifurcation. In Proceedings of the 2014 IEEE 27th International Conference on Micro Electro Mechanical Systems (MEMS), San Francisco, CA, USA, 26–30 January 2014; pp. 1285–1288.
5. Noori, N. Analysis of 2:1 Internal Resonance in MEMS Applications. Master's Dissertation, School of Mechatronic Systems Engineering, Simon Fraser University, Burnaby, BC, Canada, 2018.
6. Rhoads, J.F.; Shaw, S.W.; Turner, K.L. Nonlinear Dynamics and Its Applications in Micro- and Nanoresonators. In Proceedings of the ASME 2008 Dynamic Systems and Control Conference, Parts A and B, Ann Arbor, MI, USA, 20–22 October 2008; pp. 1509–1538.
7. Nayfeh, A.H.; Mook, D.T. *Nonlinear Oscillations*; John Wiley & Sons: Hoboken, NJ, USA, 2008.
8. Golnaraghi, F.; Behreyni, B.; Marzouk, A.; Sarrafan, A.; Lajimi, S.A.M.; Pooyanfar, O.; Noori, N. Vibratory Gyroscope Utilizing a Nonlinear Modal Interaction. U.S. Patent WO2016179698A1, 17 November 2016.
9. Lan, C.; Qin, W.; Deng, W. Energy harvesting by dynamic unstability and internal resonance for piezoelectric beam. *Appl. Phys. Lett.* **2015**, *107*, 093902. [CrossRef]
10. Ramini, A.H.; Hajjaj, A.Z.; Younis, M.I. Tunable Resonators for Nonlinear Modal Interactions. *Sci. Rep.* **2016**, *6*, 34717. [CrossRef] [PubMed]
11. Bajaj, A.K.; Chang, S.I.; Johnson, J.M. Amplitude modulated dynamics of a resonantly excited autoparametric two degree-of-freedom system. *Nonlinear Dyn.* **1994**, *5*, 433–457. [CrossRef]
12. Golnaraghi, M.F.; Moon, F.C.; Rand, R.H. Resonance in a high-speed flexible-arm robot. *Dyn. Stab. Syst.* **1989**, *4*, 169–188. [CrossRef]
13. Sarrafan, A.; Bahreyni, B.; Golnaraghi, F. Analytical modeling and experimental verification of nonlinear mode coupling in a decoupled tuning fork microresonator. *J. Microelectromech. Syst.* **2018**, *27*, 398–406. [CrossRef]
14. Sarrafan, A.; Bahreyni, B.; Golnaraghi, F. Design and characterization of microresonators simultaneously exhibiting 1/2 subharmonic and 2:1 internal resonances. In Proceedings of the 2017 19th International Conference on Solid-State Sensors, Actuators and Microsystems (TRANSDUCERS), Kaohsiung, Taiwan, 18–22 June 2017; pp. 102–105.
15. Sarrafan, A.; Bahreyni, B.; Golnaraghi, F. Development and Characterization of an H-Shaped Microresonator Exhibiting 2:1 Internal Resonance. *J. Microelectromech. Syst.* **2017**, *26*, 993–1001. [CrossRef]
16. Vyas, A. Microresonator designs based on nonlinear 1:2 internal resonance between flexural -flexural and torsional -flexural structural modes. Ph.D. Dissertation, Department of Mechanical Engineering, Purdue University, West Lafayette, Indiana, 2008.
17. Cowen, A.; Hames, G.; Monk, D.; Wilcenski, S.; Hardy, B. *SOIMUMPs^{TM} Design Handbook, Rev 8.0.*; MEMSCAP Inc.: Durham, NC, USA, 2011.

MDPI

St. Alban-Anlage 66

4052 Basel

Switzerland

Tel. +41 61 683 77 34

Fax +41 61 302 89 18

www.mdpi.com

Micromachines Editorial Office

E-mail: micromachines@mdpi.com

www.mdpi.com/journal/micromachines